Design and Analysis of Cryptographic Algorithms in Blockchain

Design and Analysis of Cryptographic Algorithms in Blockchain

Ke Huang
Yi Mu
Fatemeh Rezaeibagha
Xiaosong Zhang

CRC Press
Taylor & Francis Group
Boca Raton London New York

CRC Press is an imprint of the
Taylor & Francis Group, an **informa** business

First edition published 2021
by CRC Press
6000 Broken Sound Parkway NW, Suite 300, Boca Raton, FL 33487-2742

and by CRC Press
2 Park Square, Milton Park, Abingdon, Oxon, OX14 4RN

CRC Press is an imprint of Taylor & Francis Group, LLC

Library of Congress Cataloging-in-Publication Data

ISBN: 978-1-032-03932-9 (hbk)
ISBN: 978-1-032-03991-6 (pbk)
ISBN: 978-1-003-19012-7 (ebk)

Typeset in CR10
by KnowledgeWorks Global Ltd.

*To blockchain practitioners
and researchers.*

Contents

Foreword

The rise of crypto-currencies has driven researchers worldwide to pursue the study of blockchain. Blockchain is a cryptographic project which consists of the hash function, digital signature and other cryptographic primitives. Although cryptography has started its early military use since 170 years ago, it was until 1970 that public-key cryptography (PKC) been formally introduced as a strictly-scientific discipline. From the last 40 years, PKC schemes have been studied extensively and intensively. With the practice of complexity theory and provable security, researchers can prove the security and efficiency of PKC scheme rigorously. Although blockchain has attracted million dollar fortunes in a short time, it originated from modern cryptography. In other words, blockchain is just one of the successful implementations of many PKC schemes. Therefore, the summarization and analysis of PKC schemes help readers gain empirical insights in refining blockchain's design or replicating its success.

There have been countless research works on PKC schemes in the last 40 years. With recent advances in blockchain, researches on blockchain have grown significantly. However, there is lack of hieratical and systematic surveys or tutorials on how to design PKC schemes securely for blockchain. Particularly, we mean algorithms which are designed to cater for decentralization, immutability, trustless and any other features of blockchains. By hieratical, we mean the provision of basic knowledge and methodology to validate the design of a PKC scheme. By systematic, we mean review and summation of each PKC primitive, such as encryption, signing, hashing, signcryption etc. Meanwhile, training the reader to implement designs after systematic study is also important. Based on the above, this book intends to provide readers with lessons to securely design and analyze PKC schemes for blockchain.

This work was supported in part by the National Key R&D Program of China under Grant No.2017YFB0802300 and Grant No.2018YFB08040505; National Natural Science Foundation of China under Grants 62002048, 61872087, U19A2066; Research Initiation Fund (Central Universities) under Grant No.Y030202059018061; Blockchain Research Lab of UESTC-Chengdu Jiaozi Financial Holding Group CO.LTD; Chengdu Blockchain Industrial Innovation Center.

Preface

Blockchain is a popular application scenario for algorithmic design and implementation. Since blockchain itself is fundamentally a cryptographic project composed of independent crypto-primitives such as hash function and digital signature, the study of which is of significance to the advances of modern cryptography. Public-key cryptography (PKC) is founded by mathematic theory, complexity theory, provable security, etc. It allows the researcher to investigate a scheme rigorously prior to blockchain's success, public key cryptography (PKC) has been studied for more than 40 years. The emergence of blockchain is just one of the many successful cases in PKC history. To encourage continuous innovation and long-term theoretical research, a systematic tutorial to investigate how to design secure and practical PKC scheme for blockchain is highly important.

The study of PKC scheme is challenging for many cryptographic beginners as it involves inter-disciplines and cross-domains knowledge. Although the mathematic theories have provided the researcher with a rigorous methodology for research, they also set up a pretty high entering level for starters. Among these mathematic theories, provable security is known as the most frequently used and universally-acknowledged tool to validate a secure design of PKC scheme. In a nutshell, a reduction is often used to reduce the breaking of PKC scheme's security to the intractability of the mathematic hard problem.

To help readers study PKC and blockchain schemes systematically, this book is dedicated to the secure design and analysis of blockchain-oriented PKC schemes. Concretely, we will explain how to design and analyze PKC schemes based on literature review, case study and implementation. With three sections discussing background and basics, five sections reviewing each independent PKC primitive, one section specifically focusing on the implementation, this book is supposed to provide step-by-step guidance for a cryptographic beginner and blockchain engineers to master PKC schemes for blockchain. As an ultimate goal, we hope to encourage and inspire rigorous and longterm research for both cryptography and blockchain.

Ke Huang, Yi Mu, Fatemeh Rezaeibagha, and Xiaosong Zhang

Acknowledgments

In the completion of this book, I thank my co-authors. Professor Yi Mu has provided his continuous support of my research and his work in this book. Meanwhile, I would also like to appreciate Dr. F. Rezaeibagha who has contributed a lot of efforts to revise this work and make constructive suggestions. Meanwhile, I'd like to appreciate the supports from PhD adviser: Prof. Xiaosong Zhang for encouragement. This book could not be achieved without the academic, financial, and other supports from friends, colleagues and scientists who we share common interests and passions in blockchain and cryptographic researches.

The preparation of this book can be traced back to three years ago, when I decided to narrow my research interests from PKC study to blockchain-oriented research. This is an important turning point for me at which I started to turn my theoretic knowledge into practical use. After studying and exploiting from previous works, I planned to write a book focusing on PKC and blockchains design. More importantly, the process of writing allows me to better study and learn from previous works, and eventually become more rigorous and skilled in future research.

Introduction

Recent years have witnessed the quick rise of crypto-currencies, which makes blockchain a hot research topic as well as a successful project in which to invest. Bitcoin, the first successful crypto-currency has taken the first share of crypto-market and accumulated more than 10 billion US dollars a fortune in the last 10 years. Blockchain is fundamentally driven by using cryptographic hash functions and digital signatures, a sequence of blocks is maintained via distributed nodes in real-time without reliance on any central trusts. The success of Bitcoin has also attracted extensive researches on cryptography in search for the second Bitcoin. Public-Key Cryptography (PKC) is the most common infrastructure used in modern information infrastructure. PKC has been widely implemented in computer and network transmission. Generally, PKC is characterized by assigning each user with public and private key pair. As a result, secure communication can be achieved in a public channel. When studying the security of a proposed PKC scheme, provable security is generally practiced. Specifically, in practicing provable security theory, a security reduction is programed to reduce from the premise that a cryptographic scheme is secure to the breaking of a known mathematical hard problem. As founded by mathematics, security can be validated scientifically and universally. However, it is not easy for a researcher to program such reduction. It is not possible to finish reducing correctly without getting familiar with the inherent cryptographic design and experiencing extensive practice. Meanwhile, since each distinct PKC scheme is highly dependent on the mathematic primitive for design, there is no universal or trivial way to program security reduction. This implies the necessity of summation and extensive studies of PKC schemes, especially for those designed in blockchain scenario. In this book, we aim to investigate how each type of cryptographic primitive is designed for blockchain use. More specifically, we seek to understand how security reduction is programed based on a distinct PKC algorithmic design. Here, we specifically focus on the scheme with blockchain-related features, such as decentralization, trustless, threshold-based, etc. In a nutshell, this book is dedicated to the secure design and analysis of blockchain-based PKC schemes. We will show how to design a scheme and prove its security and applicability in blockchain. To do so, we will survey some classic PKC schemes and classify them for a separate discussion. By analyzing how security and efficiency proofs are plotted, we seek to provide readers with systematic views and experience in designing practical

PKC scheme. This will help to gain insightful knowledge in future research in either PKC scheme or blockchain.

Chapter 1

Overview

1.1 Chapter Introduction

In this chapter, we will review the background of this book. Then, we give general motivation and foundations of this book. Next, we explain basic steps to validate a proposed public-key cryptographic scheme. Finally, we will provide our major contributions of this book.

1.2 Overview of Background

1.2.1 Blockchain

The past ten years have witnessed a quick rise of crypto-currencies, which is fundamentally driven by the blockchain technology. Blockchain introduced by Nakamoto [1] is characterized as a decentralized, immutable public trust ledger. As the most successful crypto-currency, Bitcoin has received tremendous popularity in the financial market. It has accumulated more than $10 billion US dollars fortunes so far, and it has still accounted as the first share in the global crypto-currencies' market. In the last decade, researchers and technicians have actively contributed to the development of crypto-currency. As a result, blockchain technology has been widely studied and implemented. Many crypto-currency systems have emerged with variable features. Ethereum [2], a crypto-currency platform, is famous for the provision of life circle for virtual currency. Instead of transmitting value and trust, it also promotes rich and varied applications. The rise of crypto-currency industry is believed to be a driving force to modern information infrastructure, paving our way to a smarter and more advanced society.

To explain, blockchain is a chain of blocks linked by cryptographic hashes where each block consists of transactions. Further, each transaction data is authenticated by a digital signature. To note, hash function and digital signature are both primitives of cryptography. Since blockchain is fundamentally a cryptographic project, research on cryptography (especially for security) is of vital importance to blockchain's longterm prosperity.

1.2.2   Modern Cryptography

Cryptography generally refers to the study and practice of hiding sensitive information. Diffie and Hellman [3] introduced the concept of public-key cryptography (PKC), beginning the era of modern cryptography. During the last 40 years, extensive and intensive researches on PKC have been carried out, in which they can be categorized by distinct primitives, such as data integrity, authentication, encryption, etc. The idea of PKC works by negotiating secret keys via a secure channel in advance, where each user is assigned with a pair of private and public keys. As a result, the secure transmission can be achieved in a public channel afterwards. The proposal of PKC concept has inspired and encouraged researchers to pursuit research on digital signature [4], encryption [5] and many other primitives (e.g. [6–8]). On the one hand, PKC schemes are used in a non-blockchain and centralized settings, where each user entrusts a central authority (like a bank) for transactions. On the other hand, PKC schemes are also used in blockchain-based and decentralized settings (e.g. Bitcoin). Specifically, PKC-based signature schemes refer to the use of the private key to authenticate a message, while verification is fulfilled via the public key. PKC-based encryption facilitates the encryption process without prior negotiation on decryption keys. The ciphertext encrypted under a specific public key can only be decrypted via the corresponding private key. Since PKC schemes are founded by mathematics, the study of which is supposed to be strictly-scientific by practicing mathematical theories.

1.3   Motivation and Foundation of This Book

1.3.1   General Motivation

This book seeks to investigate how to design practical public-key cryptographic schemes for blockchain, which is both theoretically secure and efficient [9]. In the following, we give a general motivation for this book.

In the recent advances in crypto-currencies, more and more functional requirements are taken into considerations for design purposes. In addition to decentralization, users nowadays wish to preserve their privacy rights and enhance their security. On the one hand, these demands are caused by the frequent cyber attacks witnessed on the Internet, which caused significant loss to users and service providers. On the other hand, these newly proposed features are also strong selling points to attract more users to purchase crypto-currencies. As a result, PKC schemes have been increasingly integrated into the design of crypto-currency. For example, non-interactive zero-knowledge proof (NIZK) [6] has been used for designing Zerocoin project [10]. To explain, the NIZK is a PKC primitive to achieve secure transmission without leaking any useful information. Meanwhile, ring signature has been used to design crypto-projects, like Monero [11] and RingCT 2.0 [12] projects. To clarify, the ring signature is a PKC primitive to achieve anonymous authentication where users identities are hidden during the verification stage. But how can researchers and technicians convince users of the security of proposed PKC schemes in real practice? Provable security is an ideal answer because it is the most frequently and universally practiced theory to prove a given PKC scheme's security. Specifically, a researcher needs to define a security model which simulates how security goals should be achieved. Based on the given model, a security reduction is programed in a way that a cryptographic scheme is secure to the breaking of a known mathematic hard problem. Meanwhile, complexity theory is also applied to evaluate the cost of PKC scheme theoretically, where assessments of approximate space, time and computing costs are enabled.

1.3.2   Basic Foundations

Foundations of this book are shown in Figure 1.1. As illustrated in Figure 1.1, theories of provable security [13] and complexity [14] are the major theoretical foundations [15]. Meanwhile, blockchain serves as an application background under which PKC scheme is built. Explanations are given as below:

- **Provable Security Theory:** Introduced by Goldwasser and Micali [4], this theory formulates rigorous methodology to investigate security based on NP hard problem. In this book, it yields a security model to define security requirements according to the specific blockchain environment. Conversely, this environment yields a threat model to describe adversary's power in attacking the PKC scheme. Following these models, security reduction is programed to achieve the provable security. See more details in Section 2.4.

- **Complexity Theory:** This theory formulates mathematical format to evaluate the cost of executing a PKC scheme in the worst case. Following this theory, the designer of the PKC scheme can generate complexity proof to validate a scheme's performance. Conversely, starting from point of application, a performance criteria can be used to regulate and direct PKC's design from a complexity perspective. See more details in Section 2.3.

- **Blockchain:** Notion of blockchain introduced by Nakamoto [1] indicates decentralized and public trust ledger. Here, this notion serves as a general infrastructure, application scenario and indicators for schematic designs. To solve challenging issues in blockchain, researchers are motivated to design PKC schemes (e.g. digital signature scheme) from different perspectives. Meanwhile, to achieve trustless and distributed configurations for blockchain, particular design requirements arise (e.g. anonymity, lightweight in computation, etc).

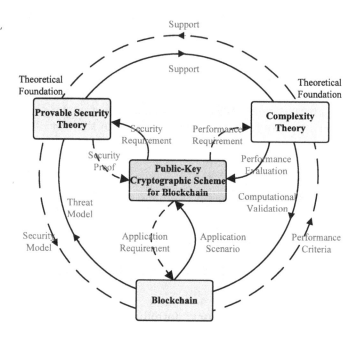

FIGURE 1.1: Foundations for PKC and Its Relationships

1.4   Overview of Designing Practical PKC Scheme

1.4.1   Validating a PKC Scheme

Early cryptography started its use in military, but rigourous security analysis is revealed only after its official use many years ago (e.g. Enigma). Goldwasser and Micali [4] introduced the notion of provable security which opens the era of modern cryptography. Based on NP hard problem, a researcher is enabled to design and validate a secure PKC scheme in a rigorous manner. In the past 40 years, provable security has been widely acknowledged as a formal and scientific methodology for PKC research.

In PKC research, it is required to validate a design by proving it is secure [16,17]. As suggested in Figure 1.2, the proposal of PKC scheme with no supporting evidence is not convincing and valid regarding its security and efficiency. When practicing provable security and complexity theories correctly (discussed in Section 1.3.2), a PKC proposal can be acknowledged by other researchers.

FIGURE 1.2: Acknowledged PKC Proposal

1.4.2 Steps to Investigate PKC Scheme

We show the necessary steps taken to design and analyze a public-key cryptographic scheme in Figure 1.3.

- In the first step, we explain our motivation to design the PKC scheme in blockchain-related application. For example, the need to encrypt or sign certain data on the chain, an encryption or signature scheme can be designed.

- Secondly, we formulate a preliminary design based on the above requirements (i.e. security, efficiency, and functionality requirements which the scheme is supposed to satisfy).

- In the third step, we complete a detailed construction based on certain mathematic parameter. The use of a mathematic parameter is highly related to a certain mathematic hard problem.

- In the fourth and fifth steps, security and efficiency analyses are provided for investigating the scheme's security and performance, respectively. In the security analysis, a reduction is programed to reduce the breaking of the scheme's security to a certain mathematic hard problem. The efficiency analysis is provided by practicing complexity theory and conducting software and hardware simulations.

- In the sixth step, conclude the practicality of a proposed cryptographic scheme based on previous analysis.

The above steps constitute all the necessary steps to validate a secure and practical design of PKC scheme for blockchain.

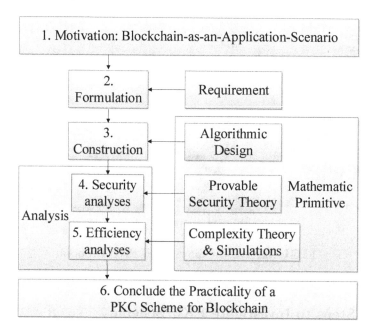

FIGURE 1.3: Steps to Investigate the PKC Scheme

1.5 Contributions

This book is dedicated to the secure design and analysis of PKC schemes for blockchain and crypto-currencies. In particular, we will show how to construct a secure PKC scheme for blockchain and prove its security formally. Concretely, we will survey the classic and blockchain-oriented PKC schemes.

With a focus on how to design a secure PKC scheme, we will explore how security analysis is formulated under the pre-defined security model. We demonstrate the main structure of this book in Figure 1.4. Following this heuristic structure, readers are given systematic views on how to design secure PKC scheme step by step. Significant contributions of this book can be summarized as follows:

In chapter 2, we systematically review the prior knowledge required for designing a PKC scheme for blockchain. Then, we start from the basics of blockchain and PKC to mathematic theories which will be used later in this book. The topic of this section includes preliminary definitions of blockchain and PKC scheme, complexity theory, provable security and numeric theory. This section is supposed to give readers the basic knowledge for following sections.

In chapter 3, we will present a survey on the technical background of crypto-currencies and blockchain projects. Specifically, we will take a glimpse at the research line of blockchain and PKC scheme in the last 10 years. By enumerating each crypto-currency appeared in blockchain history, we briefly discuss the pros and cons of its design. This helps readers to grasp the motivations and necessities of designing a successful PKC scheme.

From chapter 4 to 7, we will formally discuss how to design PKC scheme and validate its security. Our topics include digital signature, encryption, hash function and zero-knowledge proof designed under PKC infrastructure. To provide a hieratical view for readers, we will extend our discussion as follows: First, we start with the motivation for designing the PKC primitive. Then, we show the general technical information on this kind of primitive. For example, if the topic is encryption, we will introduce semantic security, which is the key security for encryption. Next, we will review some classic designs to help the reader learn the general design pattern. Then, we conduct case studies on different schemes, where security and efficiency are provided for ease of comparison. Here, the discussion on efficiency is further extended by discussing its time, space and computing cost for evaluation. Lastly, a chapter summary is provided to conclude each section.

In chapter 8, we will discuss other tools and primitives used to optimize blockchain's performance or security. The concerned knowledge is not necessarily related to PKC design, but is frequently used during research, like bloom filter, accumulator, etc. Meanwhile, we will also study concepts of consensus protocol, smart contract and other elements of blockchain . Here, we mainly give a technical overview which is not necessarily related to PKC design. More or less, this section provides the reader with useful knowledge to pursuit blockchain research.

In chapter 9, we will discuss the regulatory issue of blockchain, and show how to design secure PKC scheme to regulate blockchain based on the previous study. This section will briefly introduce motivation, policy and challenges associated with blockchain regulatory. More importantly, we show how to practically implement PKC design according to previous studies .

In chapter 10, we will conclude this book. We first give a general conclusion on the design and analysis of all PKC schemes discussed in this book. Then, we survey some new constructions and applications for PKC and blockchain. Next, we review the open problems and challenges in designing PKC schemes. Finally, we end this book with an empirical and insightful conclusion.

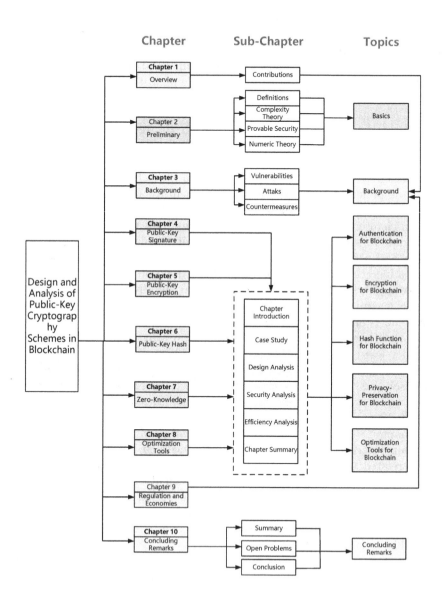

FIGURE 1.4: The Layout of This Book

1.6 Chapter Summary

This chapter gives an overview of the background, motivation and techniques of this book. It briefly introduces our primary research objects: blockchain and public-key cryptography. Then, it explains our motivation to write this book and its foundations. It details the necessary steps and techniques to validate public-key cryptographic schemes. In the end, it concludes our major contributions. To sum up, this chapter starts our journey to designing practical public-key cryptographic algorithms for blockchain. In short, it basically answers what we do, why we do it, how we do it and what's new here.

Chapter 2

Preliminaries

2.1 Chapter Introduction

In this chapter, we briefly give the basic definitions of blockchain and PKC schemes. Then, we present the basic principle of analyzing the security and efficiency of PKC schemes. At last, we give fundamental mathematical hard problems which will be used in this book.

After reading this chapter, you will be able to:

- Learn basic definitions of blockchain and PKC schemes.

- Interpret basic philosophy on how to analyze security and efficiency of PKC scheme.

- Learn fundamental mathematical hard problems which will be used in this book.

2.2 Definitions of Research Objects

2.2.1 Defining Blockchain

We formally review the basics of blockchain according to Garay's work [18]. Let $G(.)$ and $H(.)$ be cryptographic hash functions on input of $\{0,1\}^{\kappa}$. Denote $s \in \{0,1\}^{\kappa}$ as block hash, $x \in \{0,1\}^{*}$ as blockchain content, $ctr \in \mathbb{N}$ as a register. Block B is valid (described as $\mathsf{validblock}_q^D(B)$) only if $(H(ctr, G(x,s)) < D) \wedge (ctr \leq q)$ holds. Here, the parameter $D \in \mathbb{N}$ denotes a difficulty level for a block, $q \in \mathbb{N}$ denotes a bound to register ctr. Since blockchain is a chain of blocks in a sequence, the rightmost block is the head while the leftmost block is the genesis block (the first block to begin with). We define the head as $\mathsf{head}(\mathcal{C})$. Therefore, empty string is also a chain, i.e. $\mathsf{head}(\varepsilon) = \varepsilon$ where ε is the empty string. To append a new block $\mathcal{B} =< s, x, ctr >$ to a chain $\mathcal{C}$ with $\mathsf{head}(\mathcal{C}) =< s, x, ctr >$, the new chain is defined by $\mathcal{C}_{\mathsf{new}} = \mathcal{CB}$ where $\mathsf{head}(\mathcal{C}_{\mathsf{new}}) = \mathcal{B}$ and $s = H(ctr, G(s,x))$. The length of a chain $\mathcal{C}$ is defined as $\mathsf{len}(\mathcal{C}) = n$. Further, we define a vector $x_{\mathcal{C}} =< x_1, \cdots, x_n >$ and use x_i to denote the ith block in chain $\mathcal{C}$ where $i = \{1, \cdots, n\}$. Above are the basic definitions of a typical blockchain and protocol. In this book, blockchain content is generally described as $x \in \{0, 1\}^*$. Cryptographic hash functions frequently used in this book are described as $G(\cdot)$ and $H(\cdot)$. We refer readers to Garay's work [18] for more details.

2.2.2   Defining PKC Schemes

Public-Key Cryptography (PKC) refers to the application of a cryptographic mechanism to offer dome security services. Under such mechanism, each user is assigned with a private and public key pair. Generally, a private key is secretly kept by the user and public key is publicly known by everyone for the verification. Relevant security services include privacy-preservation, integrity auditing, non-repudiation, confidentiality, etc. Each service indicates a specific security goal. For example, integrity auditing generally seeks to check data integrity for a target file, like cloud-based remote data checking.

A PKC scheme indicates a detailed construction of associated algorithms to achieve a particular security service. Hereby, for each PKC scheme algorithm, a series of steps is constructed to calculate an output from the given input. Generally, an algorithm is divided into deterministic algorithms and probabilistic algorithms. The former implies one which returns the same result each time for the same instance as input. The latter indicates one which returns output differently for the same input each time. The notation (t,ϵ) is used to indicate an algorithm which outputs an answer to a given instance in time t with probability ϵ. Here, ϵ can be denoted as either probability or advantage differently, depending on the proposed scenario. For example, if an algorithm is used to calculate an answer compared to other algorithms or conditions, ϵ is called advantage; otherwise, it is called probability. Meanwhile, $t(\lambda)$ is denoted as function with an input of security parameter λ. Security parameter indicates a parameter of cryptographic primitive, such as the size of an encryption key. If $t(\lambda) = \mathcal{O}(\lambda^{n_0})$ where $n_0 > 0$, this means $t(\lambda)$ is a polynomial function. Further, we denote $\epsilon(\lambda) = \frac{1}{\mathcal{O}(\lambda^{n_0})}$ as a non-negligible advantage if $n_0 > 0$.

Formally, we say a PKC scheme is secure if no adversary can break the scheme with non-negligible advantage (i.e. $\epsilon(\lambda) = \frac{1}{\mathcal{O}(\lambda^{n_0})}$ and polynomial time (i.e. $t(\lambda) = \mathcal{O}(\lambda^{n_0})$) with security parameter λ. Further, we say an algorithm that can break a PKC scheme (or hard problem) with (t, ϵ) where t is polynomial time (i.e. $t(\lambda) = \mathcal{O}(\lambda^{n_0})$) and ϵ is non-negligible with security parameter λ, is a computationally efficient algorithm. To prove a PKC scheme is secure, a reduction is often programed by reducing the premise that a cryptographic scheme is secure to breaking of a known mathematical hard problem. In other words, in a reduction, we generally start by the premise of a computationally efficient algorithm to break a given PKC scheme. Nevertheless, due to the intractability of the underlying hard problem, no such algorithm exists as a solution. By this means, we can reduce the breaking of a PKC scheme to a mathematically hard problem. Therefore, the associated PKC scheme is provably secure.

2.3 Complexity Theory

Computational complexity [19] is the theory which bridges the gap between mathematics and theoretical computer science, offering scientific methodology to investigate cryptography. Concretely, it offers a strictly mathematical format to evaluate the cost of executing an algorithm in the worst case. PKC schemes' complexity include: time complexity, space complexity and computation complexity. Differently, they imply resources required to run algorithm from timing, storage and computing perspectives. Informally, if we define $\mathcal{A}$ as an algorithm and λ as the security parameter, the time to execute $\mathcal{A}$ is bounded by $\mathcal{O}(\lambda)$ which means that the worst time taken to run $\mathcal{A}$ is defined as $\mathcal{O}(\lambda)$. The same definition goes with space and computation complexity. In general, complexity theory gives us the basis to perform theoretic analysis on PKC schemes' performance formally and universally. More details can be found in [20].

If a random variable follows the uniform distribution and is independent of any given information, then there is no way to relate a uniformly random variable to any other information by any means of computation. This is precisely the security basis behind the only unconditionally (or information-theoretically) secure encryption scheme: one-time pad, that is, mixing a uniformly random string (called key string) with a message string in a bit by bit fashion. The need for independence between the key string and the message string requires the two strings to have the same length. Unfortunately, this poses an almost impassable limitation for practical use of the one-time-pad encryption scheme.

Nevertheless (and somewhat ironic), we are still in a "fortunate" position. At the time of writing, the computational devices and methods widely available to us (hence to code breakers) are based on a notion of computation what is not very powerful. To date, we have not been very successful in relating, via computation, between two pieces of information if one of them merely looks random while in fact, they are completely dependent on one another (for example, plaintext, ciphertext messages in many cryptosystems). As a result, modern cryptography has its security based on a so-called complexity-theoretic model. Security of such cryptosystems is conditional on various assumptions that certain problems are intractable. Here, intractable means that the widely available computational methods cannot effectively handle these problems.

We should point out that our fortunate position may only be temporary. A new and much more robust model of computation, quantum information processing (QIP), has emerged. Under this new computation model, exponentially many computation steps can be parallelized by manipulating the so-called super-position of quantum states. Consequently, many useful hard problems underlying the security of complexity-theoretic based cryptography will collapse, that is, will become useless. For example, using a quantum computer, factorization and multiplication of integers will take similar time if the integers processed have similar sizes, and hence, e.g. the famous public-key cryptosystems of Rivest, Shamir and Adleman (RSA) will be thrown out of stage. However, at the time of writing, the QIP technique is still not integrated into practical applications. The current record for factoring a composite is 15, which is the least size, odd and non-square composite integer.

To conclude, we are concerned much about the QIP for the time being. The rest of this chapter introduces our less-powerful conventional computational model and the complexity-theoretic-based approach to modern cryptography.

2.3.1   Computational Complexity Measurement

When we measure the computational complexity for an algorithm, it is often difficult or unnecessary to pinpoint the constant coefficient in an expression that bounds the complexity measure. Order notation allows us to ease the task of complexity measurement.

Definition 2.1 Order Notation: *We write $\mathcal{O}(f(n))$ to denote a function $g(n)$ such that there exists a constant $c > 0$ and a natural number N with $|g(n)| \leq c|f(n)|$ for all $n \geq N$.*

2.4   Provable Security Theory and Proof Techniques

Since public-key cryptography is founded by mathematics, PKC scheme is therefore based on mathematical primitives. While we are constructing specific PKC schemes, we rely on specific mathematical primitives which correspond to mathematic hard problems. In other words, the security of each PKC scheme is highly independent of its mathematic design. Next, we briefly introduce the concept of provable security.

Goldwasser and Micali [4] first proposed the concept of provable security, where the security model and the adversary model are proposed to evaluate security. A security model defines what security goals a PKC scheme achieve. A threat model defines the power of an adversary to attack a given scheme. Based on the above, the reduction is programed to reduce from the premise that a cryptographic scheme is secure to breaking known underlying mathematical hard problems.

When we formulate a security reduction, the following issues should be considered:

1. Whether the security model is constructed correctly to simulate the desired security goals. This reflects the possible behaviors an adversary may conduct in an actual attack scenario.

2. Whether the hard problem is properly embedded in the reduction.

3. Whether the number of queries made by the adversary during the reduction is small enough or constant.

4. Whether the adversary is limited by computations or certain conditions. For example, with the provision of a random oracle or the privilege to make queries in advance.

However, provable security and security reduction are not perfect. With the advent of quantum cryptography, the adversary's computation power will surge and exceed the description of the current security models. This will invalidate the majority of provable security analyses. Meanwhile, this framework requires full captures of potential attacks. With advances in cyber attacks and heterogeneous nature of the network, breaking the PKC scheme is becoming hard to detect and investigate. This makes a practical security model more complex to define and a provable security hard to achieve. Nevertheless, so far, the provable security theory is a universally acknowledged methodology to study a cryptographic scheme. Refer to [21] for more details.

In this book, we cover the latest development in PKC and blockchain designs. Particularly, we seek to study how security reduction is programed to PKC schemes and blockchain security. We also focus on the design and analysis of PKC schemes under the decentralized and blockchain-oriented applications. However, since the security model is highly dependent on the mathematic primitive and algorithmic design, this makes it non-trivial. We will conduct extensive reviews on classic PKC schemes as a starting point, and we will analyze the recent designs of the PKC schemes with a focus on the blockchain-based schemes. Later, we will also implement relevant designs.

Next, we will investigate several essential proving techniques. To showcase how to use these techniques to programe security analysis, we will give concrete instantiations. Here, we only focus on techniques which can be generalized and formalized. Some minor techniques such as case analysis, deduction, though is important, will not be specifically explained.

2.4.1 Proof by Security Reduction

The general methodology for provable security is to reduce an alleged attack on a cryptographic scheme to a solution to a mathematically intractable hard problem. The simulator runs a reduction following the defined reduction algorithm. A security reduction introduces how to generate a simulated scheme and how to reduce an attack on this scheme to solve an underlying hard problem. The desired reduction to contradiction proof must have the following properties:

1. The reduction should be efficient and tight [22]. An inefficient reduction, even if it is in polynomial time, may provide no practical merits at all between an attack and a solution to a hard problem.

2. The intractability assumptions which are required for a scheme being secure should be as weak as possible. Since weak assumptions are easier to satisfy using more practical and available cryptographic constructions. Consequently, cryptographic systems using weaker assumptions provide higher security confidence than those using stronger assumptions.

2.4.2   Proof by Sequences of Games

To prove security using the sequence-of-games approach (also known as game hopping), one proceeds as follows. One constructs a sequence of games (a.k.a, game hopping), Game 0, Game 1, $\cdots$, Game n, where Game 0 is the original attack game with respect to a given adversary and cryptographic primitive. Let S_0 be the event S, and for $i = 1, \cdots, n$, the construction defines an event S_i in Game i, in which it is related to the definition of S. The proof shows that $\Pr[S_i]$ is negligibly close to $\Pr[S_{i+1}]$ for $i = 0, \cdots, n-1$ and that $\Pr[S_n]$ is equal (or negligibly close) to the target probability. From this, and the fact that n is a constant, it follows that $\Pr[S]$ is negligibly close to the target probability, and security is proved.

To instantiate, game hopping is a method for proving the security of a cryptographic scheme. In a game hopping proof, we observe that an attacker running in a particular attack environment has an unknown probability of success. We then slowly alter the attack environment until the attackers success probability can be computed [23,24]. This book introduces two well-known types of game hop: transitions based on indistinguishability, Transitions based on failure events.

2.4.2.1   Transitions based on Indistinguishability

In such a transition, a small change is made to the step, where, if detected by the adversary, it would imply an efficient method of distinguishing between two distributions that are indistinguishable.

For example, suppose P1 and P2 are assumed to be computationally indistinguishable distributions. To prove that $|\Pr[S_i]|\Pr[S_{i+1}]|$ is negligible, one argues that there exists a distinguishing algorithm D that "interpolates" between Game i and Game $i+1$, so that when given an element drawn from the distribution P_1 as input, D outputs 1 with probability $\Pr[S_i]$, and when given an element drawn from the distribution P_2 as input, D outputs 1 with probability $\Pr[S_{i+1}]$. The indistinguishability assumption then implies that $|\Pr[S_i]|\Pr[S_{i+1}]|$ is negligible.

2.4.2.2 Transitions based on Failure Events

In such a transition, one argues that Games i and $i+1$ proceed identically unless a certain failure event F occurs. Specifically, define two games on the same underlying probability space, the only differences between them is the rules for computing certain random variables. Hence, when we say that the two games proceed identically unless F occurs is equivalent to say tha

$$S_i \wedge \neg F \iff S_{i+1} \wedge \neg F. \tag{2.1}$$

That is, the events $S_i \wedge \neg F$ and $S_{i+1} \wedge \neg F$ are the same. If this is true, then we can use the following fact, which is completely trivial, yet is frequently used in these types of proofs:

Lemma 2.1 *(Difference Lemma). Let A, B, F be events defined in some probability distribution, and suppose that $A \wedge \neg F \iff B \wedge \neg F$. Then $|\Pr[A] - \Pr[B]| \le \Pr[F]$*
Proof. As is a simple calculation, we have

$$|\Pr[A] - \Pr[B]| = |\Pr[A \wedge F] + \Pr[A \wedge \neg F] - \Pr[B \wedge F] - \Pr[B \wedge \neg F]|$$

$$= |\Pr[A \wedge F] - \Pr[B \wedge F]|$$

$$\le \Pr[F]. \tag{2.2}$$

Note that the second equality follows from the assumption that $A \wedge \neg F \iff B \wedge \neg F$, so in particular, $\Pr[A \wedge \neg F] = \Pr[B \wedge \neg F]$. The final inequality follows from the fact that both $\Pr[A \wedge F]$ and $\Pr[B \wedge F]$ are numbers between 0 and $\Pr[F]$.

To prove that $\Pr[S_i]$ is negligibly close to $\Pr[S_{i+1}]$, it suffices to prove that $\Pr[F]$ is negligible. Sometimes, this is done using a security assumption (i.e. when F occurs, the adversary has found a collision in a hash function, or forged a MAC), while at other times, it can be done using a purely information-theoretic argument.

Basically, the event F is defined and analyzed in terms of one of the two adjacent games' random variables. The choice is arbitrary, but typically, one of the games will be more suitable than the other in terms of allowing a concrete proof.

2.4.3   Proof by Contradiction

Proof by contradiction is adopted to prove that a proposed scheme is secure against any adversary in polynomial time with non-negligible advantage: We have a mathematical problem that is believed to be hard. We assume an adversary can break the proposed scheme and prove that the adversarys attack can be reduced to solve the underlying hard problem by a security reduction. We conclude that the proposed scheme is secure because the underlying hard problem becomes easy if the breaking assumption is true [21].

A proof by contradiction is described as follows.

> A mathematical problem is believed to be hard.
> If a proposed scheme is insecure, we prove that this problem is easy.
> The assumption is then false, and the scheme is secure.

The proof by contradiction for public-key cryptography is explained as follows. Firstly, we have a mathematical problem that is believed to be hard. Then, we give a breaking assumption that there exists an adversary who can break the proposed scheme in polynomial time with non-negligible advantage. The adversary is assumed to be able to break the proposed scheme by following the steps described in the proof by testing. Next, we show that this mathematical hard problem is easy because such an adversary exists. The contradiction then indicates that the breaking assumption must be false. That is, the scheme is secure and cannot be broken, and therefore the proposed scheme is secure.

The contradiction occurs if and only if we can efficiently solve an underlying hard problem with adversary's help. If the underlying hard problem is easy, or we cannot efficiently solve the underlying hard problem, the proof will fail to achieve a contradiction. A proof without contradiction does not mean that the proposed scheme is insecure but that it is not provably secure. As a result, the given proof cannot convince us that the proposed scheme is provably secure.

2.4.4   Proof by Theorem

Proof by theorem [25] deals with cryptographic protocols by front manner. It is an expert system based on human empirical knowledge and skills for efficiently modeling and analyzing protocol's security. To explain, this method first formalizes the cryptographic protocol at the beginning. Then, it starts to model and produce reduction based on the formalized protocol. However, this method suffers from an incomplete definition of adversary's model (unavailability of a whole picture of adversary's malicious behavior). Meanwhile, it mainly tackles with passive attacks and treats cryptographic algorithms by blackbox. Consequently, it cannot detect vulnerabilities that resulted from the illogical design of cryptographic algorithms. Proof by theorem does not require any physical equipment for testing, and therefore it executes in fast and verifiable way. Massive researchers generally use it as an ideal theoretical method for validating cryptographic schemes.

2.4.5 Random Oracle Model

Random oracle model (ROM) is an ideal device which offers response services on queries for cryptographic use. It plays an essential role in the design of a provable secure scheme [26]. A ROM H projects n-bits string into m-bits string and satisfies the following properties: public accessibility, randomization and unpredictability, secrecy and consistency. The inner structure of ROM is unpredictable to external users. The only view acquired by an external user is the queries and responses. Based on the use of ROM in security reduction for provable security, encryption schemes can be proven in either by ROM and the standard model. In 1993, Bellare and Rogaway [26] formalized the design pattern in ROM. In short, one needs to first design the scheme in ROM and then give security reduction based on the intractable problems while assuming random oracle existence. Second, replace the random oracle model with the standard hash function H', e.g. SHA-1.

2.4.6 Standard Model

Knowing that hash function does not necessarily yield completely randomized output and cannot guarantee the security of cryptographic scheme in a realistic environment. There exists cases [27,28] of provable security scheme in ROM but not secure in the real environment. Unlike ROM, the standard model relies on no theoretical assumption such as random oracle. For example, when adopting hash function, it only exploits standard property like collision-resistance. Thus, adversary's capability is only constrained by computation time and space. It is challenging to devise secure cryptographic scheme in the standard model. Although suffering from low efficiency, the proposed scheme in the standard model is more convincing and practically secure than one in ROM. Refer to ref. [17] for more details.

2.5 Algebraic Notions and Numeric Theory

In this section, we review algebraic Notions and numeric theories used in this book.

2.5.1   Group, Ring and Fields

2.5.1.1   Group

Definition 2.2 (Group) *Let G be a non-empty set. Denote $*$ as any binary operations in G for associated elements, i.e. $* : G \times G \to G$. Here, G is a group if following conditions are satisfied:*

1. *For any $a, b \in G$, $a * b \in G$. (Closure Axiom)*

2. *For any $a, b, c \in G$, $a * (b * c) = (a * b) * c$. (Associativity Axiom)*

3. *For each group G, there exists an identity element 1_G such that $1_G * a = a * 1_G = a$ for any $a \in G$. (Identity Axiom)*

4. *For each group G, there exists an inverse $1/a = a^{-1}$ for every $a \in G$ such that $a * a^{-1} = a^{-1} * a = 1_G$. (Inverse Axiom)*

Denote a group as $< G, * >$. The $< G, * >$ is an abelian group if $a * b = b * a$ for any $a, b \in G$. Denote $|G|$ as the order of group G. If $|G|$ is finite, $< G, * >$ is a finite group. Additionally, a^k means a is multiplied k times by itself and $a^{-k} = \left(a^{-1}\right)^k$.

Definition 2.3 (Order of Group) *Let G be a group and $a \in G$ be an element. The order of the element a is the least positive integer $i \in \mathbb{N}$ satisfying $a^i = 1_G$. Denote $\mathsf{ord}(a)$ as the order of the group element. If i does not exist, call a as an element of infinite order.*

Definition 2.4 (Cyclic Group and Its Generator) *A group is cyclic if there exists an element $a \in G$ such that for any $b \in G$, there always exists $i \geq 0$ such that $b = a^i$. Therefore, a cyclic group G is formed by n repeated operations $a^1, a^2, \cdots, a^n$ for some element $a \in G$. Element a is called the generater of cyclic group G, written as: $G = < a >$. a is also called a primitive root of group's identity element 1_G.*

2.5.1.2   Ring

Definition 2.5 (Ring) *Let R be a ring, which is a set with two operations: addition $+$ and multiplication $\times$. Here, R is a ring if the following conditions are satisfied:*

1. *Under addition $+$, R is an abelian group with additive identity $\mathbf{0}$;*

2. *Under multiplication $\times$, R satisfies Closure Axiom, Associativity Axiom and Identity Axiom as specified in Definition 2.2. Denote $\mathbf{1}$ as the multiplicative identity.*

3. *For any $a, b \in R$, $a \times b = b \times a$. (Communicative Axiom)*

4. *For any $a, b, c \in R$, $a \times (b + c) = a \times b + a \times c$. (Distribution Axiom)*

2.5.1.3 Field

Definition 2.6 (Field) *Let R be a ring. If the non-zero elements in ring R forms a group under multiplication (i.e. $\times$ operation), then R is a field.*

Definition 2.7 (Finite Field) *Let q be a prime number. The finite field $\mathbb{F}_q$, is comprised of the set of integers $\{0, 1, 2, \cdots, q-1\}$ with the following arithmetic operations:*

1. *Addition: If $a, b \in \mathbb{F}_q$, then $a + b = r$ where r is the remainder when $a + b$ is divided by q and $0 \le r \le q - 1$. This is known as addition modulo q.*

2. *Multiplication: If $a, b \in \mathbb{F}_q$, then $a \times b = s$ where s is the remainder when $a \times b$ is divided by q and $0 \le s \le q - 1$. This is known as multiplication modulo q.*

3. *Inversion: If a is a non-zero element in $\mathbb{F}_q$, the inverse of a modulo q, denoted as a^{-1}, is the unique integer $c \in \mathbb{F}_q$ for which $a \times c = 1$.*

2.5.1.4 Bilinear Pairing

Definition 2.8 (Bilinear Pairing) *$\hat{e}$ is said to be a bilinear pairing if it is constructed from mapping a pair of group elements to one specific element in the group. Let G_1, G_2, G_T be cyclic groups of order q. Let g_1 and g_2 be generators of G_1 and G_2 respectively. $\hat{e} : G_1 \times G_2 \to G_T$ is a bilinear pairing if following properties are satisfied:*

1. *$\hat{e}(a^x, b^y) = \hat{e}(a, b)^{xy}$ for all $a \in G_1$, $b \in G_2$ and $x, y \in Z_q$.*

2. *$\hat{e}(g_1, g_2) \ne 1_{G_T}$ where 1_{G_T} is the identity element of group G_T.*

3. *The computation of $\hat{e}(a, b)$ is supposed to be efficient, i.e. can be executed within polynomial time $t(\lambda) = \mathcal{O}(\lambda^{n_0})$ where $n_0 > 0$ for all $a \in G_1$ and $b \in G_2$.*

Moreover, bilinear pairing can be classified in three types: Type 1, Type 2, and Type 3. Specifically, Type 1 pairing means $G_1 = G_2$; Type 2 pairing means $G_1 \ne G_2$ which there exists an efficiently one way computable homomorphism $\pi : G_2 \to G_1$; Type 3 means $G_1 \ne G_2$ which no efficiently computable homomorphism appears from G_1 to G_2 or vice versa.

2.5.1.5 Proof of Knowledge

Proof of knowledge (PoK) of a discrete logarithm enables a prover $\mathcal{P}$ to convince the verifier $\mathcal{V}$ that he knows $X = \log_g^Y$ without exposing X. By applying a Schnorr signature [29], a PoK for Y's discrete logarithm is computed as: $PoK(Y) = (c, s) \in \{0, 1\}^k \times Z_q$ where $c = H(g, Y, g^s Y^c)$. Here, the prover $\mathcal{P}$ selects a randomness $r \xleftarrow{R} Z_q$ and computes $c = H(g, Y, g^r)$ where

$H : \{0,1\}^* \to \{0,1\}^k$ is a secure and collision-resistant hash function. Then, the verifier $\mathcal{V}$ checks and accepts if $c = H(g, Y, g^s Y^c)$.

Proof of knowledge for the equality of two discrete logarithms (say, u and v with respect to the base g_0 and g_1) allows a prover $\mathcal{P}$ to convince the verifier $\mathcal{V}$ that $\log_{g_0}^u = \log_{g_1}^v$. By applying the Chaum and Pedersen [30]'s idea, the PoK for the equality of u and v's discrete logarithms is computed as $PoK(u, v) = (c, s) \in \{0,1\}^k \times Z_q$, where $c = H(g_0, g_1, u, v, g_0^s u^c, g_1^s v^c)$ and $H : \{0,1\}^* \to \{0,1\}^k$ is a secure and collision-resistant hash function. Similarly, the prover $\mathcal{P}$ selects a randomness $r \xleftarrow{R} Z_q$ and computes $c = H(g_0, g_1, u, v, g_0^s u^c, g_1^s v^c)$. Then, the verifier $\mathcal{V}$ checks and accepts if $c = H(g_0, g_1, u, v, g_0^s u^c, g_1^s v^c)$.

2.5.2   Numeric Theory

Application use of mathematical hard problems in security reduction enable researchers to prove PKC scheme's security via strictly-scientific methodology. The following assumptions are supposed to be without efficient algorithm as a solution.

Definition 2.9 (Discrete Logarithm (DL) Assumption) *DL   problem (DLP) is defined as follows: Given $(g_1, y) \in G_1^2$, compute $x \in Z_q$ such that $y = g_1^x \in G_1$. If there is no algorithm to compute DL problem in probabilistic polynomial time with non-negligible probability, then DL assumption holds in G_1.*

Definition 2.10 (Computational Diffie-Hellman (CDH) Assumption [3]) CDH problem is defined as follows: Given $(g_1, g_1^a, g_1^b) \in G_1^3$, compute $g_1^{ab} \in G_1$. If there is no algorithm to compute CDH problem in probabilistic polynomial time with non-negligible probability, then CDH assumption holds in G_1.

Definition 2.11 (Decisional   Diffie-Hellman   (DDH)   Assumption) DDH problem in G_2 is defined as follows: Given $(g_2, g_2^a, g_2^b, g_2^c) \in G_2^4$ determining whether or not $c = ab$. If there is no algorithm to compute DDH problem in probabilistic polynomial time with non-negligible probability, then CDH assumption holds in G_2.

A gap Diffie-Hellman (GDH) problem is for groups where the CDH problem is hard, and DDH problem is easy. It can be constructed from supersingular elliptic curves or hyperelliptic curves over finite fields [31].

Definition 2.12 (External Diffie-Hellman (XDH) Assumption) *XDH problem is defined as follows: Let G_1, G_2, G_T be groups of prime order p with a non-degenerate bilinear map $\hat{e} : G_1 \times G_2 \to G_T$. Determining DDH problem is hard in one of the groups G_1 or G_2 [32].*

Definition 2.13 (Symmetric External Diffie-Hellman (SXDH) Assumption) SXDH problem is defined as follows: Let G_1, G_2, G_T be groups of prime order p with a non-degenerate bilinear map $\hat{e} : G_1 \times G_2 \to G_T$. Determining DDH problem is hard in both G_1 and G_2.

Definition 2.14 (q-Strong Diffie-Hellman (q-SDH) Assumption)
q-SDH problem in G is defined as follows: Given $(g, g^a, \cdots, g^{a^q}) \in G^{q+2}$, compute $(c, g^{1/(a+c)})$. If there is no algorithm to compute q-SDH problem in probabilistic polynomial time with non-negligible probability, then q-SDH assumption holds in G.

Definition 2.15 (Bilinear Diffie-Hellman (BDH) Assumption)
BDH problem in $\mathbb{G}_1$ and $\mathbb{G}_2$ is defined as follows: Let $\mathbb{G}_1, \mathbb{G}_2$ be two groups of prime order q. Let $\hat{e} : \mathbb{G}_1 \times \mathbb{G}_1 \to \mathbb{G}_2$ be an admissible bilinear map and let P be a generator of $\mathbb{G}_1$. Given $< P, aP, bP, cP >$ for some $a, b, c \in \mathbb{Z}_q^*$, compute $W = \hat{e}(P, P)^{abc} \in \mathbb{G}_2$. If there is no algorithm to compute BDH problem in probabilistic polynomial time with non-negligible probability, then wBIDH assumption holds.

Definition 2.16 (Bilinear Inverse Diffie-Hellman (BIDH) Assumption) BIDH problem in G and G_2 is defined as follows: Given $(g_1, g_1^a, g_1^b) \in G_1^3$, $g_2 \in G_2$, compute $\hat{e}(g_1, g_2)^{c/a}$. If there is no algorithm to compute BIDH problem in probabilistic polynomial time with non-negligible probability, then BIDH assumption holds.

As we have seen above, elliptic curve points are additive groups to finite fields that are defined over multiplicative groups. The notations used for implementing and discussing schemes over elliptic curve groups differ from the notation used in some algorithms. We will informally define the most commonly used cryptographic assumptions using the additive notation of elliptic curve arithmetic, and then formally explore the assumption that the scheme we use depends on. All definitions are defined concerning an adversary $\mathcal{A}$.

Definition 2.17 (Elliptic Curve Discrete Logarithm (ECDL) Assumption) ECDL problem is defined as follows: Given $Y, P \in E(\mathbb{F}_q)$, find $x \in \mathbb{Z}$ such that $Y = x \cdot P$. If there is no algorithm to compute ECDL problem in probabilistic polynomial time with non-negligible probability, then ECDL assumption holds in $E(\mathbb{F}_q)$.

Definition 2.18 (Elliptic Curve CDH (EC-CDH) Assumption) EC-CDH assumption is defined as follows: Given $(a \cdot P, b \cdot P) \in E(\mathbb{F}_q)$ where $a, b \in \mathbb{Z}$ are unknown random numbers, compute $ab \cdot P$. If there is no algorithm to compute EC-CDH problem in probabilistic polynomial time with non-negligible probability, then we say that ECDL assumption holds in $E(\mathbb{F}_q)$.

Definition 2.19 (Elliptic Curve DDH (EC-DDH) Assumption) EC-DDH assumption is defined as follows: Given $(a \cdot P, b \cdot P, c \cdot P) \in E(\mathbb{F}_q)$ where $a, b, c \in \mathbb{Z}$ are unknown random numbers, compute $ab \cdot P \in E(\mathbb{F}_q)$. Decide whether $c \cdot P = ab \cdot P$. If there is no algorithm to compute EC-DDH problem in probabilistic polynomial time with non-negligible probability, then EC-DDH assumption holds in $E(\mathbb{F}_q)$.

2.6   Chapter Summary

This chapter review some preliminary knowledge and formal definitions for blockchain and cryptographic scheme. It lays the foundation of our learning path towards secure and efficient designs of blockchain-based algorithms. Concretely, this chapter includes definitions of research objects, complexity theory, provable security and techniques, algebraic notions and numeric theory. To clarify, definitions provide a basic language for us to describe our research objects. Complexity theory and provable security provide us with the techniques to analyze proposed schemes' efficiency and security. Algebraic and numeric theory gives us the foundation to practice complexity theory and provable security. Equipped with the above knowledge and techniques, researcher learn basics to perform schematic analysis in a strictly-mathematical manner. After reading this chapter, the reader is able to: (1). Learn basic definitions of blockchain and PKC schemes. (2). Interpret basic philosophy on how to analyze security and efficiency of PKC scheme. (3). Learn basic mathematical hard problems which will be used in the rest of this book.

Chapter 3

Background

3.1 Chapter Introduction

As an old saying goes, "Every problem has a different solution". To design practical algorithm for blockchain, our priority is to know the problems associated with blockchain. In other words, which part of blockchain needs improvement. Based on this, we can start our design and analysis. This chapter provides a systematic survey on the blockchain, associated problems and countermeasures. As surveyed in this chapter, blockchain vulnerabilities are found at system-level, software-level, algorithm-level, and data-level. The practice of cryptography helps build more sophisticated blockchains and improve early blockchains at the algorithmic level. However, there still exists problems and doubts in current blockchains. This motivates us to carry out practical design and analysis on the cryptographic algorithms designed in blockchains and as overlays to blockchains.

After reading this chapter, the reader will be able to:

- Understand mainstream blockchains and its characters.

- Enumerate vulnerabilities, potential attacks and countermeasures for a given mainstream blockchain.

- Know basically how to design practical cryptographic algorithms from a practical angle.

3.2 Introduction to Blockchain

Blockchain is a decentralized and append-only public ledger where each block is sequentially linked by cryptographic hash and consists of transactions committed by cryptographic digital signature. The first well-known blockchain prototype is Bitcoin proposed by Satoshi Nakamoto [1], a decentralized digital cryptocurrency. Bitcoin enables payments to be completed among users without any pre-established third party (e.g. bank, government, etc.). However, with further intensive research, blockchain have extended to diverse applications far beyond cryptocurrencies. For example, blockchain is used to achieve Internet-of-Things [33], Edge Computing [34], Smart City [35], Artificial Intelligence [36], etc.

A typical blockchain system has following characters:

- Decentralization: In traditional economic infrastructure (e.g. e-commerce system), a central authority (e.g. the bank or a financial sector) is responsible for validating each transaction. In addition to huge costs to maintain such central reliance, it also leads to a performance bottleneck and single point of failure. Differently, blockchain breaks the old circle by bringing decentralization feature where every node in the network is equal in power and identity. Consequently, any two users can conduct transactions directly in a peer-to-peer manner (P2P). This is guaranteed by a series of cryptographic primitives, such as hash function, digital signature, consensus protocol, etc. This reduces the huge costs of maintaining a centralized system and shifts the recurring costs to each node's network.

- Anonymity: In a public blockchain (e.g. Bitcoin), users contact each other with a designated public address. This address is generated in advance and used as pseudonymity to protect the user's actual identity. Early blockchain relies on this pseudonymity solely to achieve anonymity. However, as later discussed in [37], an attacker can deduce the user's real identity from pseudonymity easily. For example, as empirical knowledge, multiple outputs in one transaction mostly indicates an identical set of users. Other useful methods are social engineering, cross-reference check, etc. Therefore, achieving users' full anonymity is challenging and non-trivial work in blockchain. One successful blockchain project: Monero [38] attracts users by its strong anonymity preservation.

- Tamper-Resistance: Despite various extensions (public, consortium or private blockchain), blockchain is generally known as an immutable (or uneditable) trust-layer. Specifically, by using the cryptographic hash function, a hash value is computed and stored in each block header which links to the previous block. Due to the intractability of finding a hash collision (aka collision resistance hash function), it is hard to forge another hash value which outputs the same hash value. To complete a transaction, the sender and receiver need to be part of public-key infrastructure. Users use the digital signature scheme for authentication as negotiated. Due to the unforgeability of digital signature, it is hard to forge a signature and pass the verification for a transaction to which the signature is committed. Thus, transactional immutability is achieved. For public blockchain, remote history is immutable since it is both technically and economically infeasible to reverse. For the recent history (e.g. the next 5 or 6 blocks), although some minor forks may occur (by rare chance), they will be eventually discarded once most miners keep following a longer chain.

- Auditability: In Bitcoin blockchain where each block hash is linked by previous one and each transaction is committed by the digital signature and enumerated to generate a Merkle root, anyone can run verification algorithm for hashing, signing and Merkle hash to verify the validity. Further, since the above verification proceeds with time stamp, each node in the blockchain network can maintain a consistent version of blockchain and avoid a single point of failure. Also, it helps achieve traceability and transparency [39].

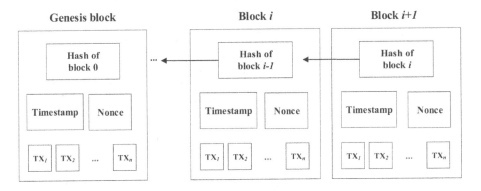

FIGURE 3.1: Bitcoin Blockchain Structure

3.2.1 Bitcoin

Satoshi Nakamoto [1] proposed Bitcoin in 2008 as the first blockchain. Then, Bitcoin was later implemented in 2009 and achieved a considerable success in academia and industry in the past decade. By the year of 2020, Bitcoin has reached more than 19 billion dollars in capital market. As Bitcoin blockchain is the prototype of many blockchains, we briefly introduce it in the next.

As shown in Figure 3.1, the Bitcoin blockchain is characterized by its linear chain structure where each block is sequentially chained by time stamp via a cryptographic hash. Specifically, each block is linked to the previous one by a hash pointer. The first block in the chain is called genesis block. In block body, there are a number of transactions $\{TX_1, \cdots, TX_n\}$ committed by digital signature (as we will discuss in Section 4). The timestamp is used to prove that data existed at the time, and each timestamp incorporates the previous one to form a chain. Most importantly, a consensus mechanism called: Proof-of-Work [1] is employed to implement the above algorithms in a distributed network via peer-to-peer basis. In short, PoW dictates global timestamp without reliance on any central trust by one-CPU-one-vote. The majority of nodes (i.e. 51% of nodes in the network) determines the longest chain. It is economically infeasible for an attacker to reverse or split the chain for bad. We refer readers to ref. [1] for more details on PoW.

A block comprises the *block header* and the *block body*. As illustrated in Table 3.1, the *block header* consists of block version, previous block's hash (parent block hash), Merkle tree root's hash, timestamp, difficulty and nonce. As given in Table 3.2, the *block body* consists of magic number, block size, block header, transaction counter and transactions data.

TABLE 3.1:   Bitcoin block head

Item	Explanation	Size (Byte)
Version	Block version	4
hashPrevBlock	Previous block's hash	32
hashMerkleRoot	Merkle root's hash	32
Time	Current time stamp (sec)	4
Bits	Current block's hardness	4
Nonce	Randomness (from 0 to 32)	4

TABLE 3.2:   Bitcoin block body

Item	Explanation	Size (Byte)
Magic No	A Constant number	4
Blocksize	Blocksize in bytes	4
blockheader	6 data items	80
Transaction counter	A positive integer	1-9
Transactions	Transaction information	$m * n \ (n > 250)$

Generally, there are five steps to confirm a transaction in the Bitcoin blockchain, the workflow is given in Figure 3.2. We briefly elaborate the procedures as follows:

1. The user broadcasts a signed transaction in the network. For a given time period, the miner pack and verify the transactions. After that, the miners try to solve a mathematic puzzle. This process relates to a consensus mechanism where nodes in the network complete to write and maintain the blockchain in inconsistency.

2. Once the miner manages to break the puzzle with an answer (nonce), he packs the transactions in a block together with the derived answer to the puzzle. He broadcasts the block over the network in order to record it on the chain.

3. On receiving the new block, other miners can append the new block in the chain's tail and keep mining on top of it. Corresponding reward (incentive) is given to the miner who breaks the puzzle.

4. While miners run proof-of-work to compete to write the new block on chain, this block is initially confirmed by blockchain system by a timestamp.

5. After a while (say, 10 minutes), the block is officially recorded on-chain and confirmed by the majority of nodes. Typically, it is difficult to reverse a block when at most 5 or 6 blocks were appended after it. Though by rare chance, when a fork occurs, the block can be reversed. However, it is impossible to reverse a block in a remote history (when at least 6 blocks were appended after it).

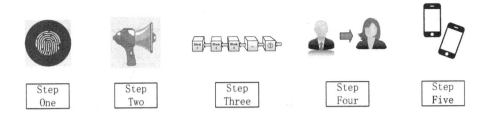

FIGURE 3.2: Bitcoin Blockchain Payment Flow

3.2.2   Ethereum

Ethereum [2] is an exceptional paradigm and platform for blockchain 2.0 (Bitcoin stands for blockchain 1.0). It is featured by an open and decentralized platform supporting Decentralized Applications (DApps) atop the blockchain. The introduction of multiple primitives (smart contracts, proof-of-stake, etc.) brings rich applications and possibilities to blockchain life circle. For ease of comparison, we highlight four major differences between Ethereum (blockchain 2.0) and Bitcoin (blockchain 1.0) as follows:

1. Bitcoin blockchain is based on the transaction-centered model, called unspent transaction outputs (UTXO). UTXO is the core principle to process (generate and verify) transaction. Every legally spent transaction is traceable and it links to one or several outputs. As a result, a sequence of payment is formed, where the head is the mining reward, and the tail is the unspent output. Unlike UTXO, Ethereum adopts account-based model to support DApps. This model displays user's balance of inherent cryptocurrency (Ether) directly. This paradigm is easier to be searched and programed.

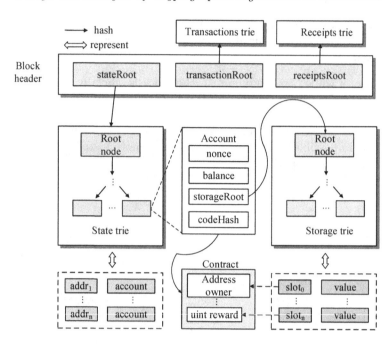

FIGURE 3.3: Instantiation of Ethereum Blockchain

Source: Data from Chen et al. [40], A Survey on Ethereum Systems Security:
Vulnerabilities, Attacks, and Defenses. *ACM Computing Surveys (CSUR)*, 53(3): 1–43,
2020.

2. Ethereum has been revolutionized by introducing smart contracts in 2015, breaks the boundary of Bitcoin for being a decentralized cryptocurrency only. As the smart contract is Turing-complete, it basically can be used to achieve an arbitrary computer program. However, since blockchain is generally irreversible, it is hard to fix the security vulnerability in smart contract once the contract has been deployed. Another problem is the secure programing languages, and tools to achieve secure and sound smart contracts.

3. Ethereum adopts proof-of-stake (PoS) rather than PoW as a consensus protocol to reach an agreement. Unlike one-CPU-as-a-vote, PoS determines votes according the holding time and proportion of cryptocurrencies owned by one node. Precisely, the PoS algorithm scales down the mining difficulty based on above factors, which helps find the answer to the puzzle quickly.

4. Ethereum charges payer a processing fee for each transaction, called gas. Unlike the mining fees rewarded to miner for solving the PoW puzzle in the Bitcoin blockchain, the gas fee is mandatory for each payer who

launches the transaction. Simply, charing the gas is for preventing a bus of network resources resulted from infinite loops due to launching transactions. The payer will need to pay the gas before launching the transaction. Meanwhile, each operation during the transaction will consume some gas. Insufficient gas will trigger a system alarm which will undo previous operations and return all consumed gas to the miner. Recently, extensive and intensive researches have been carried out to study and cope with gas-related problems found in Ethereum.

In Ethereum, there are two types of accounts: *externally owned accounts* (EOA) and *contract accounts*. EOA keeps a user's funds and is linked by a public key. The private key corresponds to the public key used to access EOA, which the user keeps secretly. Contract account stores smart contract in executable bytecode to enable autonomous and smart applications. An EOA or contract account has a dynamic state. The state of a blockchain is defined by the states of the accounts on the blockchain. Specifically, the state is determined by following parameters:

- *nonce*, used to track the number of transactions initiated by the EOA's owner. It can also be used to track the number of contracts held by the contract account.

- *balance*, denotes the amount of money in the EOA or contract account. It is converted into the smallest sub denominations of Ether: Wei (i.e. 10^{-18} Ether).

- *storageRoot*, denotes the hash root of the accounts storage data structure *trie*. *trie* is used to record a contracts state variables.

- *codeHash*, denotes the hash value of a contract accounts bytecode.

The physical design of Ethereum blockchain is similar to Bitcoin blockchain: a chain of blocks linked by cryptographic hash. Meanwhile, the payment workflow, which shows how a transaction is processed in such infrastructure from generation to confirmation, is also similar (as depicted in Figure 3.2). However, the inherent technical philosophy and parameters included are significantly different. Ethereum is a state machine for transaction. This machine is the only one in the world but existed in a distributed manner around the world. The execution of a transaction alters the states of the accounts (EOA and contract account). So, the state of the blockchain (world state) is altered as well. The world state reflects the status of all accounts in the Ethereum system. For ease of discussion, we give an instantiation in Figure 3.3.

To derive the world state, a special type of data structure is used : Merkle tree (also known as: Merkle trie). The root of Merkle trie, generated from iterative hashing of two leaf nodes, is used to verify all leaf nodes' integrity and structure in the tree. For each block's header of each block in Ethereum blockchain, three hash roots are recorded: *stateRoot*, *transactionRoot* (records transaction data) and *receiptsRoot* (records the execution data of transactions). The block body contains a set of transactions just as Bitcoin blockchain. Differently, the transaction is specified by following parameters: *nonce* (a counter), *recipient, value, input, gasPrice*, (v, r, s) (a tuple for Elliptic Curve Digital Signature Algorithm, ECDSA). To facilitate the search, the Merkle trie stores $(key, value)$ pairs such that the path from the root to a leaf node corresponds to a *key* and the leaf node includes a *value*. For more details, refer to ref. [40].

3.2.3 Other Cryptocurrencies

Bitcoin is the defacto standard for many newly proposed cryptocurrencies. Many cryptocurrencies are either directly descended from Bitcoin (e.g. Namecoin [41]) or built on top of the blockchain network [42]. These cryptocurrencies are collectively known as Altcoins [37]. We summarize some well-known Altcoins in Table 3.3 together with a brief introduction. Specifically, we categorize them by cryptography-centered and non-cryptography-centered Altcoins by leveraging the inherent use of new cryptographic primitives (e.g. ring signature, zero-knowledge proof) in comparison with Bitcoin.

3.2.3.1 Cryptographic Scheme-Centered Altcoins

Miers et al. [45] proposed *Zerocoin* which exploits zero knowledge proofs [55] to offer anonymity for users. Before spending a zerocoin, the user is required to generate a commitment. This commitment will be recorded in blockchain for verification. Later, to spend a zerocoin, the user must generate a zero knowledge proof that he owns an equal value of Bitcoin. By using of zero knowledge proof, no leakage of useful information is possible. As a result, others can not associate the user with the spent transactions. Unlike mixing technique which works as an overlay, zerocoin seeks to provide anonymity via inherent cryptographic primitives. The drawback is that the zerocoin involves complex cryptographic protocol which leads to a bottleneck of entire system.

Ben-Sasson et al. [46] extended *Zerocoin* to *Zerocash* (also called: Zcash) by exploiting a practical version of zero knowledge proof: Zero-Knowledge Succinct Non-Interactive Argument of Knowledge (zk-SNARK). Additionally, strong anonymity is achieved which preserves users' transaction amount as well as recipient's addresses. However, it assumes a trusted authority to generate secrete parameters for zk-SNARKs during the setup stage. This is a challenging problem [56].

TABLE 3.3: Review of altcoins

Altcoin	Techniques	Remark
Cryptography-centered Altcoins		
Byzcoin [43]	Collective signing	Achieve low consensus latency via micorblocks and collective signing.
Monero (*XMR* [44])	LRS	Adopt ring signature to achieve fully anonymous cryptocurrency.
Zerocoin [45]	NIZK & Commitment	Extend Bitcoin introducing NIZK for anonymity protection.
Zerocash [46]	zk-SNARK	Propose fully decentralized and anonymous payment scheme by leveraging zk-SNARK.
Other Altcoins		
Bitcoin (*BTC*) [1]	Pseudonym	The first successful decentralized cryptocurrency.
Bytecoin (*BCN*) [47]	CryptoNote	Adopt CryptoNote to achieve strong anonymity for transactions.
Counterparty (*XCP*) [48]	decentralized platform	Decentralized and open platform built on Bitcoin protocol.
Dash (*USDT*) [49]	Mixing	Require user's interaction to enable mixing to blind the transaction.
Ethereum (*ETH*) [2]	Smart contract	Achieve decentralized platform for rich and autonomous applications.
Litecoin (*LTC*) [50]	Scrypt	Adopt scrypt to avoid centralized mining power, but induces heavy computations.
Libra [51]	BFT consensus	Introduce unpopular but so-called "secure" language Rust and assume most nodes are honest during consensus.
Mastercoin (*MSC*) [52]	Colored coin	Adopt proof-of-authenticity, support transaction revocation and escro fund for security.
MimbleWimble [53]	Confidential transaction	Improve Bitcoin's privacy and scalability, rely on cryptographic primitive.
Namecoin [41]	Decentralized DNS	Designed to decentralized power of domain management and avoid single point of failure.
Ripple (*XRP*) [54]	Byzantine agreement	Adopt new Byzantine agreement protocol for reaching agreement.

BFT: Byzantine Fault Tolerance; NIZK: Non-interactive zero-knowledge; LRS: Linkable Ring Signature; DNS: Domain Name System; zk-SNARK: Zero-knowledge succinct non-interactive argument of knowledge.

Monero [44] is a notable cryptocurrency with a strong focus on privacy. By utilizing ring signature and confidential address, payer and payee's identity privacy is guaranteed simultaneously. Specifically, each user is assigned with an additional key pair, i.e. there are two public keys (public view key, public send key) and two private keys (private view key, private send key) for each user. The sender generates a one time public key by taking inputs of two public keys. Then, he uses the derived one-time public key to generate a one-time address for the recipient (known as stealth address) to achieve address anonymity. Besides, a ring confidential transaction protocol is implemented to ensure transaction anonymity. The drawback is that monero is much larger than Bitcoin in its block size. This is resulted from involving numerous cryptographic operations during the protocol.

Byzcoin [43] is a Bitcoin-like cryptocurrency with efficient consensus and high transaction processing speed. To explain, it is implemented with scalable byzantine fault-tolerant consensus and collective signing as cryptographic solutions.

3.2.3.2   Other Altcoins

Jedusor propose *MimbleWimble* [53] to enable confidential transactions (CT). As CTs can be integrated to reclaim space, mimblewimble is known for its scalability. The drawbacks are the vulnerability to DoS attacks and not supporting the smart contract.

Bytecoin [47] is a cryptocurrency with unlinkable and untraceable transactions. It implements CryptoNote protocol which binds CPU to GPU for mining during PoW consensus to enable fair mining competition.

Ripple [54] implements a novel Byzantine agreement consensus algorithm with low-latency and robustness. This cryptocurrency is fast in confirmation and energy-friendly. However, it is not a fully decentralized framework that also suffers from traditional byzantine agreement consensus threats.

Counterparty [48] is known for its naive cryptocurrency (XCP) generated by destroying Bitcoins by Proof-of-Burn (PoB). The philosophy of PoB is to "burn" Bitcoins proportionately to the XCP consumed by the corresponding user. The intention is to reward counterparty's developers with a fraction of Bitcoins for processing transactions. To ensure Bitcoin is indeed burned during the transaction, relevant protocols are introduced to validate the legitimacy.

Duffield et al. [49] proposed *Dash* (formerly known as Darkcoin) as a privacy-focused cryptocurrency based on Bitcoin. They proposed an improved proof-of-work consensus algorithm which uses a chain of hashing algorithms as replacement of SHA-256. Their goal was to quickly respond to fluctuations of a large mining pool.

Litecoin [50] implements scrypt (an encryption algorithm) in mining algorithm. In addition, it allows users to participate in mining via ordinary GPU (consumer-level) and achieve fast transaction confirmation (approx. 2.5 minutes). It is scalable, anonymous and cheaper to deploy among users.

Libra [51] is novel cryptocurrency launched by Libra Association (lead by Facebook). It seeks to become a stable cryptocurrency leveraged by real purchasing power (PPP). It seeks to achieve a cryptographically authenticated data structure with libra protocol running on top of it from a technical view. It is known as the Libra Reserve concept where the Libra Association tends to manage libra cryptocurrencies for good (e.g. investment, foundation, etc). Libra is similar to mainstream cryptocurrency in design.

Mastercoin [52] adopts enhanced Bitcoin core and proof-of-authenticity as a consensus algorithm. It is designed without altering the Bitcoin groundwork or generating a substitute technology for taking care of new rules. It is easy to deploy and provide services for complex financial tasks.

Namecoin [41] originates from the study of decentralized namespaces. While DNS is an important and centralized component of the Internet, researches seek to break the reliance on central system and avoid single-point-of-failure, resulting in a significant loss. Namecoin is designed to decentralize the power of domain management, which is a namespace used to register name/value pairs in blockchain for individuals to trade.

3.3 Security Vulnerabilities, Attacks & Countermeasures

3.3.1 Vulnerabilities of Altcoins

Since blockchain has been widely deployed, its corresponding system, networks, software, algorithm are suffering from a variety of external threats. For early Bitcoin blockchain [1], due to the anonymous, freely-accessible natures of network and lack of financial surveillance, Bitcoins are exploited by malicious users. As frequently reported by mass medium [57], Bitcoins are used to finance terrorists, black market trades, etc. Due to Bitcoin circle simplicity, its influence and disruption are restricted by ever-increasing financial regulations. However, in an era of blockchain 2.0 started by Ethereum [2], diverse applications and intricate architecture are introduced to blockchain. Meanwhile, more potential vulnerabilities are introduced to blockchain from various aspects. One typical example is the ever-increasing and serious security breaches in smart contract witnessed in recent study [58]. As given in Figure 3.4, we divide the blockchain architecture into six layers. Then, we list the concrete attack targeting specific blockchain layer and Altcoin in Table 3.4. Based on the above, we discuss the general vulnerabilities of blockchain from the bottom layer to the top layer (in Figure 3.4) as follows:

- **Data Layer Threat:** This layer describes the threat targeting the cryptographic primitives used in blockchain, such as hash function, Merkle tree, signing scheme and encryption scheme. More detailed definitions for threats are given in the first category of Table 3.4.

- **Network Layer Threat:** This layer describes the threat targeting the P2P network and relevant procedures during which the blocks are supposed to be transmitted and verified. Relevant attacks are: Sybil attack [59], eclipse attack [60] and rooting attacks.

- **Consensus Layer Threat:** This layer describes the threat targeting the consensus mechanism based on which each mutually-untrusted node reach an agreement in the blockchain network. Relevant attacks are: selfish mining [58], block discarding, and block withholding [61] attacks.

- **Incentive Layer Threat:** This layer describes the threat targeting the incentive mechanism associated with the operations and encouragement of blockchain mining, e.g. the price of gas and mining reward for the miner. The relevant attack is the abuse of unfair gas price [58,62].

- **Contract Layer Threat:** This layer describes the threat targeting smart contract. For example, abuse of unfair gas price to exhaust system resources or infiltrate through the backdoor of problematic function for disruption. Relevant attacks are: DoS attack [58] and unauthorized code execution [63].

- **Application Layer Threat:** This layer describes the threat targeting blockchain-based application. Application-oriented attacks are profound since attacks may exploit any unnoticed loophole or tiny bug. In other words, one or several layers as mentioned earlier of attacks are possible to be launched. What is more, instead of a network attack, the attacker may use side-channel attack [64] to infiltrate the blockchain system or the device physically by various methods.

3.3.2   Privacy Threat of Altcoins

Privacy refers to sensitive information about a person (or company), such as personal financial state and real identity behind pseudonymity. Since blockchain is increasingly involved in people's social life, some privacy are hidden among the ever-increasing data recorded in blockchain and Altcoins. The breach of users' privacy will not likely to impose an imminent threat to Altcoin's security but will result in the financial or political loss to user directly (or indirectly). For example, early Bitcoin blockchain relies on pseudonyms as a means to preserve privacy. It uses multiple addresses as an identity for the for transaction. However, as pointed out by a study [37], user's actual identity and financial status can be easily revealed by cross-reference comparison and empirical analysis through social engineering. The leakage and misuse of sensitive information could cause significant loss to users [65]. Therefore, the privacy threat is a critical problem for the Altcoin. However, nowadays, privacy can be easily breached due to easily-accessible network and increasing-powerful analysis techniques. Following levels of leakage can summarize privacy threats.

Network-Layer Privacy Threat: Revelation of users' real identity from pseudonymity is considered as a significant privacy leakage at this stage. An attacker can eavesdrop or actively collect communicated data between neighbor nodes via available network techniques. This allows him to analyze the topology path to locate the original sending node of a transaction. Suppose sender's and recipient's addresses are openly recorded (e.g. like bitocin), the attacker can thus link the associated pseudonymities with real identities.

Transaction-Layer Privacy Threat: Extracting the user's identity or financial information are significant threats at this stage. For Bitcoin, the transaction is publicly verifiable. An attacker can analyze the transaction log to derive user's financial information, such as the balance of personal account, currency flow, trade pattern, etc. Identity privacy is also at the risk of exposure. The attacker can deduce the user's real identity from pseudonymity by analyzing associated transactions, for example, based on the empirical knowledge that multiple outputs in one transaction mostly indicates an identical set of users (e.g. multiple miners are assigned to the same mining pool). Another example could be an account that regularly receives large amounts of Bitcoins with zero output, which is supposed to be a saving account for a company or corporate.

Application-Level Privacy-Leakage: Users' misbehaviors in operating blockchain services result in privacy leakage, for instance, a user adopts unique pseudonymity for each website. This provides an attacker with the means and chances to guess his real identity by a survey and correlation. Meanwhile, the type of application or infrastructure under which blockchain is implemented provides attackers with particular approaches to derive privacy. For instance, regarding smart grid-based blockchain, the attacker can extract users power consumption data from blockchain log to guess his financial capabilities, living habits, absence hours from home, etc.

3.3.3 Attacks of Altcoins

We list the concrete attacks against Altcoins and the resulted consequences in Table 3.4. Based on Table 3.4, we further discuss them from cryptographic perspective and non-cryptographic perspectives. The former type of attack is supposed to be theoretically hard, e.g. breaking unforgeability of the signature, collision resistance of hashing, etc.

TABLE 3.4: Threats and Attacks against Altcoins

Attack	Target	Consequences
Data layer from cryptographic perspective		
Key-exposure [67]	User's private key	Steal user's private key to access his wallet and transact on behalf.
Signature forgery [71]	User's signature	Forge to spend coin on behalf of the real user.
Crack ciphertext [72]	User's ciphertext	Breach the encryption scheme to derive sensitive information or manipulate blockchain.
Collision attack [73]	User's hash function	Break hash function or find a collision to manipulate blockchain's mining process.
Attacks from non-cryptographic perspective		
Data layer		
Double spending [74]	User's transaction	Cheat to spend coin more than once in rapid concession.
Unfair gas attack [58,62]	User's money	Exploit user's money or waste network resources from inappropriate gas price.
Network layer		
Sybil attack [59]	Blockchain network	Control multiple pseudonyms to compromise the network collaboratively.
Eclipse & Rooting attack [60]	Specific user(s)	Block and control the communication of specific node(s).
Consensus layer		
Selfish mining & Goldfinger attack [58]	Honest miners	Abuse forking mechanism to influence and control mining (Goldfinger attack).
Contract layer		
DoS attack [58]	Smart contract & network	Launch intense attacks to exhaust system or network renounces in short time.
Unauthorized code execution [63]	Smart contract	Infiltrate through inappropriate design of function to abuse code execution for bad.

DoS: Denial of Service; NIZK: Non-Interactive Zero-Knowledge; LRS: Linkable Ring Signature.

We argue that although current popular cryptographic primitives (e.g. PKI-based signing and encryption schemes) and their underlying cryptographic assumptions (e.g. discrete logarithm assumption) [3] are sound (NP-hard). However, there is no guarantee that it will still be in the near future when quantum computer and its theory continue to breakthrough. There are still some flawed and problematic designs that turned out to be a weak primitives based on which cryptocurrency is built [66]. For example, key exposure [67] generally refers to a security threat which leads to the exposure of user's private key. It has been identified in chameleon hash [67], public-key signing [68] and encryption [69] schemes. For an Altcoin which relies on the cryptographic scheme with key exposure, the attacker may seek to steal user's private key by launching exposure attack (as defined by the corresponding security model). Once the attacker successfully obtains the private key, he can use it to access user's digital wallet and spend the Altcoin as he wishes.

From a non-cryptographic perspective, attacks can be launched from various aspects (as discussed in Figure 3.4). These attacks are much easier to deploy than breaking the underlying cryptographic primitives. Many attacks have been witnessed, and some of them are still efficient and practical to launch [42]. Although assuming half of the network nodes are honest is a strong premise, the researcher has already discovered some newly proposed chains have been manipulated by large pool, controlling over 61% computing power of the entire network. Particularly, DoS attack [58] tends to be a practical attack with no satisfied countermeasure so far.

To summarize, Altcoins suffer a broad domain of attacks, including system, network, algorithm, data, etc. The data layer and contract layer are the most vulnerable sectors as data and contract are irreversible once recorded and deployed [58]. Specifically, academia is now paying great attention to the defence and security of smart contract [58]. Cryptography provides sound, and rigorous primitive to solve smart contract security [70]. In the next part, we will introduce cryptographic primitives as countermeasures to solve security problems in Altcoins.

3.3.4 Countermeasures

In this part, we try to give countermeasures for security and privacy vulnerabilities (and attacks) against Altcoins. Specifically, we give answers from cryptographic and non-cryptographic perspectives. To start, we first introduce four cryptographic schemes which are broadly used as primitive countermeasure.

- **Zero-Knowledge Proof (ZKP) and Its Variants:** Goldwasser et al. [55] proposed the notion of zero-knowledge proof. Zero-knowledge proof (ZKP) protocol allows a prover to convince a verifier of knowing a secret via a statement while the verifier learns nothing about the secret. Non-interactive zero-knowledge proof (NIZK) is a variant of ZKP where the

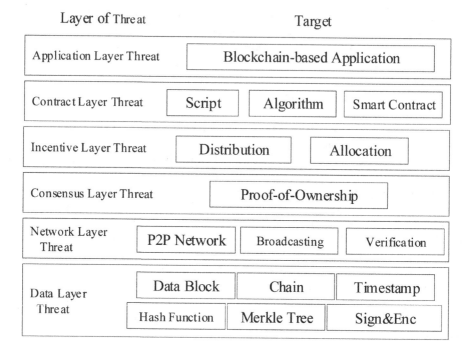

FIGURE 3.4: Architectural Threats against Altcoin

proof can be independently verified by anyone once generated without interactions. The zero-knowledge succinct non-interactive argument of knowledge (zk-SNARK) is a practical variant of ZKP. It produces argument with small and constant size while formalizing security by assuming adversary with bounded computational power. Here, the difference between argument and proof relates to the assumption of adversary's power. ZKP and its variants are building blocks for many cryptographic paradigms, such as privacy-preserving encryption, signing, etc. Mainstream ZKP variants like zk-SNARK suffer from centralized key generation. It relies on an honest generator's assumption to output the common reference string (a parameter returned in the key generation). Although some Altcoins partially solved this problem by introducing multi-party computation, computing expenditures rise significantly with the number of participants involved. Meanwhile, solving this scalability problem is not trivial.

- **Anonymous Signature (AS):** Anonymous signature allows a signer to authenticate a message without exposing his identity to the verifier, such as group signature and ring signature. Chaum and Van Heyst [75] introduced group signature, which allows a group manager to set up a group of users so that any group members can sign on behalf of the whole group. Rivest et al. [76] proposed notion of ring signature, it erases the need for a group manager which greatly simplifies the setup phase for signing. Unlike the group signature, the signer can randomly choose a set of users' public keys to form a ring in which his public key is hidden. Since such a ring is formed in spontaneously, other users are unaware that they are summoned for generating the signature. As no central trust is required, it achieves unconditional anonymity. However, it also raises the problem in tracing dispute signature since no central trust is involved. As a partial solution, linkable ring signature [77] was proposed to detect ring signatures produced by the same signer through linkability of tags via computations as an ideal primitive to detect double-spending in cryptocurrencies. Besides, designing a constant and fixed signature size is a challenge.

- **Secure Multi-Party Computation (SMPC) and Its Variants:** Andrew Yao [78] proposed secure two-party computation and Goldreich et al. [79] generalized it to multi-party computation. MPC protocol allows multiple parties to jointly compute a value over their private inputs while the adversary learns nothing about each secret input but the output. SMPC is a protocol to ensure secure and complicated operations between trustless parties. Secret-sharing-based MPC (SS-MPC) relies on oblivious transfer (OT) to obscure the participant's view about the protocol's ending cycle. As a result, it is hard for each participant to cheat and exploit others' privacy during collaborative computations. According to the different assumptions of colluded parties and how they are colluded (statically or adaptively), SMPC protocols vary in their designs and security. So far, MPC has been successfully implemented in achieving e-lottery, e-voting, e-auction, etc. It has been actively studied and used to design cryptocurrencies [80] and other projects (e.g. Enigma [81]). However, it suffers from scalability and inefficiency problems when introduced in the blockchain. Since blockchain is free to join and attracts many users, complexity costs generated from SMPC rise dramatically when users surges.

- **Fully Homomorphic Encryption** Homomorphic encryption (HE) is a special type of encryption which allows a third party to manipulate encrypted data without decryption. Since direct computation is permitted on encrypted data without violating decryption, both data operability and privacy features are guaranteed. HE can be interpreted as a non-interactive version of secure multi-party computation since no one is required to collaborate to generate the output. Rivest et al. [82] proposed the first fully homomorphic encryption (FHE) in 1978, in which it supports both addition and multiplication operations on encrypted data. It was attractive as it was still after 30 years ago that Gentry [83] proposed the first plausible and practical FHE. Despite continuous studies, current FHE schemes still cannot support real-time application. To tackle this, optimizing works start from various perspectives, including software, hardware and schematic optimization. Also, even though proposed FHE satisfy specific security features, it is suggested to achieve IND-CCA1 security for general purpose applications.

Based on the given vulnerabilities in Table 3.4 and primitive countermeasures, we give countermeasures for various attacks against Altcoins in Table 3.5. As before, we divide countermeasures to cryptographic and non-cryptographic perspectives. We argue that quantum cryptography [84] is a generic solution for traditional public key infrastructure (PKI-based) schemes in that quantum computer and theories improve the computing power, which makes many computationally hard problems (e.g. Discrete logarithm problem) easy to solve. This, in turn, threaten the foundation of many PKI-based primitives and schemes.

To cope with the key exposure problem [67], the key update is only a naive and trivial solution. The main reason is that the leakage problem is not ultimately solved by periodically updating or assigning a new private key to the user. It only cures the symptoms, not the disease. Thus, this calls for a schematic enhancement which satisfies the security model for key-exposure freeness. For instance, the key-exposure problem was identified by Ateniese et al. [67] in the early chameleon hash scheme. They solved this problem by enhancing chameleon hash in ref. [85] (Chaum Pederson Commitment) with a customized identity attached to each chameleon hash output. Similarly, a collision attack is another attack scenario that threatens Altcoin's security built on hash function. Generally, for a symmetric hash function, it is computationally hard to derive a hash collision [73]. However, for asymmetric (public key) hash function (e.g. chameleon hash), it is efficient to use the private key to generate a hash collision. The collision attack means that an attacker efficiently derives hash collision without using the corresponding private key. Again, the solving problem depends on solving the schematic problem in the hash scheme's specific design. Camenisch et al. [86] achieved collision-resilience by enhancing their scheme with the ephemeral trapdoor (generated by private key). For signature forgery and cracking ciphertext, FHE and SMPC are possible and generic solutions instead of schematic enhancement (that only targets for one specific scheme). The reason is, FHE offers secure manipulation of signature or ciphertext which offers additional security on signature (or ciphertext) than traditional signing or encryption schemes. SMPC also impedes the attacker to generate signature or decrypt ciphertext by himself. This eliminates one possible way for the attacker to launch attack or learning useful information.

From non-cryptographic perspectives, each attack's countermeasure against specific Altcoin is too specific and not much common. This is because each kind of attack is launched from various aspects (different layers) and targets for different areas or roles (user, miner or the entire system). Meanwhile, the use of single protocol or tool does not likely to be helpful since attacks can be launched in a combined or consecutive manner (or both). This also

H]

TABLE 3.5: Countermeasure for vulnerabilities and attacks

Attack	Countermeasure	Explanation
Countermeasure against cryptographic vulnerabilities		
Key-exposure [67]	Key update & Schematic enhancement	Periodically update key, deactivate leaked key and achieve zero leakage of key's info.
Signature forgery [71]	Schematic enhancement & QC & FHE & SMPC	It depends on concrete threats and security models.
Crack ciphertext [72]	Schematic enhancement & QC & FHE & SMPC	It depends on concrete threats and security models.
Collision attack [73]	Schematic enhancement & QC	Design secure hash function with strong collision-resistance.
Attack	Countermeasure	Remark
Countermeasure against practical attacks		
Data layer		
Double spending [74]	Linkability & Revocation & Malleable signature	Propose double spending Detection (LRS [77]) and revocable transaction [87].
Unfair gas price [58]	Gas price adjustment	Design a mechanism to dynamically adjust the gas price to maintain the fairness and soundness in blockchain system [62].
Network layer		
Sybil attack [59]	Xim [88]	Use multi-round mixing with unlinkability to thwart sybil attacks.
Eclipse & Rooting attack [60]	White lists	Identify or disable unknown incoming communications.
Consensus layer		
Selfish mining & Goldfinger attack [58]	ZeroBlock [89] & timestamp-based technique [90]	Propose solution against selfish mining by neglecting unpublished blocks for mining.
Contract layer		
DoS attack [58]	Proof-of-Activity [91]	Decentralize the power that synchronizes the transactions.
Unauthorized code execution [63]	Hard fork & formal verification [92]	Fork the problematic chain to avoid contract flaws.

QC: Quantum Cryptography; LRS: Linkable Ring Signature; NIZK: Non-Interactive Zero Knowledge; SMPC: Secure Multi Party Computing; Fully Homomorphic Encryption (FHE).

calls for a comprehensive consideration to preserve the interests of each participant (user, miner, validator, etc); otherwise, it deteriorate the basic design of Altcoin discourages the users. So, given countermeasure to attacks against altcoin from non-cryptographic perspectives are not a trivial work due to both security and efficiency factor in the availability of countermeasure.

3.4   Chapter Summary

Before diving deeply into the study of cryptographic algorithms, we should learn the problems in current blockchains and the existing problems can be solved using cryptographic algorithms. This gives the beginner a good starting point and direction for future research. This chapter first reviewed some general background of blockchain. Then, it summarized vulnerabilities, threats, attacks and countermeasures associated with various blockchains. This chapter initially categorized cryptographic algorithms as solutions to blockchain problems by four types: public-key signature, encryption, hashing and zero-knowledge proof. Following this category, this book will provide hieratical and in-depth guidance towards a practical algorithmic design for blockchain. After reading this chapter, the reader should be able to: (1) Understand mainstream blockchains and its characters. (2) Enumerate vulnerabilities, potential attacks and countermeasures for a given mainstream blockchain. (3) Know basically how to design practical cryptographic algorithms from a practical angle.

Chapter 4

Public-Key Signature Scheme for Blockchain

4.1 Chapter Introduction

To demystify the design and analysis of public key signature scheme in blockchain, this chapter offers detailed lecture for beginners by case analysis. This chapter begins with a short introduction to PKS. Then, it reviews the construction and security of several schemes: ECDSA, BLS, and MlSAG. This chapter also shows our work as a sample to help the reader take a baby step to devise a secure PKS scheme.

After reading this chapter, you will to:

- Know the formal definition of the public key signature scheme and its security model.

- Sketch the basic design of ECDSA, BLS, and MlSAG schemes.

- Understand the instantiated proving techniques.

4.2 Overview of PKS

In this section, we first present the background of PKS. Then, we review the current PKSs used in blockchain. Next, we give formal definition for PKS.

4.2.1 Introduction to PKS

There are two famous categorizationof public-key signature, one is RSA paradigm [93] based on factorization of large integer problem, and the other is ElGamal paradigm [94] based on discrete logarithm problem (DLP, as specified in Definition 2.9). DLP can be further divided to DLP over finite fields or elliptic curves (as specified in Definition 2.17). Miller [95] first introduced the elliptic curve (EC) to the public-key cryptosystem. Since that, the EC have been widely known and implemented for its outstanding efficiency among other schemes under the same security. For example, an EC-based scheme with a key length of 160 bits is as secure as RSA scheme 512 bits.

Current PKS schemes can be classified according to the intractable mathematical hard problems which provide the basis of their security:

1. **Integer Factorization (IF) based schemes** build the security based on the intractability of the IF problem, for example, RSA [93] and Rabin [96] schemes.

2. **Discrete Logarithm (DL) based schemes** build the security based on the intractability of the discrete logarithm problem (DLP) in a finite field, for example, ElGamal [94], Schnorr [97], DSA, and Nyberg-Rueppel [98,99] schemes.

3. **Elliptic Curve (EC) schemes** build the security based on the intractability of the elliptic curve discrete logarithm problem (ECDLP), for example, ECDSA [100] given in Section 4.3.

4.2.2   PKS in Blockchain

4.2.2.1   Use Case 1: ECDSA in Bitcoin

Blockchain systems such as Bitcoin [1] and Ethereum [2] use the elliptic curve digital signature algorithm (ECDSA) for the authentication. ECDSA was first proposed in 1992 by Scott Vanstone [100], which can be interpreted as the elliptic curve (EC) analogue of the DSA. As discussed in Section 4.3, ECDSA can be described as a signature in the form of $(r; s)$ (as later sketched Figure 4.1). The advantages of using EC groups rather than finite fields are:

- Efficiency and high processing speed of EC arithmetic.

- Absence of any sub-exponential time algorithms to solve ECDLP (as specified in Definition 2.17).

The ECDLP in the group $E(\mathbb{F}_q)$ is generally believed to be more difficult than the DLP in finite fields of size q [101].

4.2.2.2   Use Case 2: MLSAG in Monero

Noether and Mackenzie [102] proposed a Multi-layered Linkable Spontaneous Anonymous Group (MLSAG) signature to achieve hidden amounts for any transactions in Monero [44]. This proposal was mainly devoted to addressing the vulnerabilities that resulted from open transaction amounts in CryptNote protocol. The motivation of MLSAG is to achieve hidden amounts, origins and destinations of transactions with a practical design of group signature and relevant protocols. To note, Monero is a cryptocurrency with strong privacy-preservation to achieve fully anonymous payment. Like all anonymous cryptocurrencies, they generally adopted variants of group signature to achieve anonymous authentication. We will formalize the definitions of group signature in Section 4.5.2.

4.2.3 Definition of PKS

A PKS is a fundamental tool in cryptography to achieve authentication. The scenario of authentication can be described as follows:

- Alice wishes to convince all other parties that she publishes a message m. She first generates a public/secret key pair (pk, sk) and publishes pk to all verifiers.

- Alice digitally signs m with sk to derive a signature σ_m.

- On receiving (m, σ_m), any receiver can verify the signature σ_m with pk and confirm the origin of the message m.

A PKS scheme consists of the following four algorithms.

$\mathsf{SysGen_{PKS}}$: This algorithm takes as input a security parameter λ. It returns the system parameters SP.

$\mathsf{KeyGen_{PKS}}$: This algorithm takes as input the system parameters SP. It returns a public/secret key pair (pk, sk).

$\mathsf{Sign_{PKS}}$: This algorithm takes as input a message m, the secret key sk, and system parameters SP. It returns a signature σ_m.

$\mathsf{Verify_{PKS}}$: This algorithm takes as input a message signature pair (m, σ_m), the public key pk, and system parameters SP. It returns "1" if σ_m is a valid signature of m signed with sk; otherwise, it returns "0".

A PKS scheme's validation generally consists of correctness and security analysis based on security requirement (or security model). We generalize them as follows:

Correctness. Given any (pk, sk, m, σ_m), if σ_m is a valid signature of m signed with sk, the verification algorithm on (m, σ_m, pk) will return "accept".

Security. Without the secret key sk, it is hard for any probabilistic polynomial time (PPT) adversary to forge a valid signature σ_m on a new message $m!`$ that can pass the signature verification.

The security of a PKS scheme is modeled by running a game among a challenger and an adversary. During the interaction between them, the challenger generates a signature scheme, and the adversary attempts to break the scheme. Unforgeability is basic security for PKS scheme, which captures the signature forgery by a malicious user. Specifically, existential unforgeability against chosen-message attacks(EU-CMA) asks that it is hard for any probabilistic polynomial time (PPT) adversary (say, $\mathcal{A}$) to forge a valid signature σ_m on a new message m that can pass the signature verification and have not been queried before. To instantiate, we formalize the security model of existential unforgeability against chosen-message attacks (EU-CMA) as follows.

Setup. The challenger runs KeyGen to generate system parameters SP and a key pair (pk, sk). It sends pk to the adversary $\mathcal{A}$. The challenger keeps sk privately, and sends SP and pk to $\mathcal{A}$. The challenger uses sk to generate signature responses to queries from $\mathcal{A}$.

Query. The adversary $\mathcal{A}$ makes signature queries adaptively on messages chosen by itself. For a signature query on the message m_i, the challenger runs Sign algorithm to compute σ_{m_i} and then sends it to $\mathcal{A}$.

Forgery. The adversary $\mathcal{A}$ returns a forged signature σ_{m^*} on some m^* and wins the game if

- σ_{m^*} is a valid signature of m^*.

- Message m^* has not been queried in the query phase.

To finalize security analysis, it remains to bound the adversary $\mathcal{A}$'s probability in winning the above the game. The advantage ϵ of winning above game is the probability of returning a valid forged signature. To clarify, probability does not necessarily equal to advantage. One should also consider successful forgeries from random guesses. We define a general security requirement for PKS scheme by following definition:

Definition 4.1 (EU-CMA) *A signature scheme is (t, q_s, e)-secure in the EU-CMA security model if there exists no adversary who can win the above game in time t with an advantage ϵ after it has made q_s signature queries.*

4.3 Case Analysis: ECDSA

In this section, we review the basic construction of ECDSA and analyse its construction and security.

The Elliptic Curve Digital Signature Algorithm (ECDSA) is the elliptic curve analogue of the DSA, which was first proposed in 1992 by Scott Vanstone [100]. The motivations of using EC groups are the superiorities in efficiency and speed of EC arithmetic. To explain, elliptic curves are defined over finite algebraic structures such as finite fields. A finite field consists of a finite set of elements F together with two binary operations on F, called addition and multiplication, that satisfy certain arithmetic properties. The order of a finite field is the number of elements in the field. Most standards restrict the underlying finite field's order to be an odd prime. We refer the reader to Section 2.5.1.3 and [103] for more details.

The ECDSA scheme $\mathcal{GS} = (\mathsf{GKeyGen}, \mathsf{GSign}, \mathsf{GVer}, \mathsf{Open})$ includes four polynomial-time algorithms. A sketch of ECDSA's construction is given in Figure 4.1.

4.3.1 Construction of ECDSA

Ecliptic Curve Digital Signature Algorithm (ECDSA)
System parameters: param$_{ECDSA}$ = $\{q, FR, a, b, G, n, h\}$. Denote E as an elliptic curve over a finite field $\mathbb{F}_q$ of characteristic p and base point $G \in E(\mathbb{F}_q)$. Public key: $pk = Q$; Private key: $sk = d$.

Signer	Verifier
1. Pick a random integer k such that $1 \leq k \leq n - 1$.	8. Check whether r and s are integers in the interval $[1, n - 1]$;
2. Compute $kG = (x_1, y_1)$ and convert x_1 to an integer $\overline{x_1}$	9. Compute $e = H(m)$.
3. Compute $r = x_1 \mod n$. If $r = 0$ then return to step 1.	10. Compute $w = s^{-1} \mod n$.
4. Compute $k^{-1} \mod n$.	11. Compute $u_1 = ew \mod n$ and $u_2 = rw \mod n$.
5. Compute $e = H(m)$ where e is a bit string and $H()$ is SHA-1.	12. Compute $X = u_1 G + u_2 Q$.
6. Compute $s = k^{-1}(e + dr) \mod n$. If $s = 0$ then go to step 1.	13. Convert the x-coordinate x_1 of X to an integer x_1. Compute $v = x_1 \mod n$. Accept the signature $\sigma(m)$ if $v = r$; otherwise, reject.
7. Output signature $\sigma(m) = (r, s)$.	

FIGURE 4.1: Sketch of ECDSA Scheme

We describe ECDSA scheme as follows:

- ECDSA-Setup(param$_{ECDSA}$ $\xleftarrow{\$}$ λ): On input a security parameter λ, this randomized algorithm generates some domain parameters param$_{ECDSA}$ = $\{q, FR, a, b, G, n, h\}$ for an elliptic curve E. Output system parameters param$_{ECDSA}$ = $\{q, FR, a, b, G, n, h\}$. Here, E denotes an elliptic curve over a finite field $\mathbb{F}_q$ of characteristic p and base point $G \in E(\mathbb{F}_q)$. q is the field size of $\mathbb{F}_q$. FR is an identifier used to indicate elements of $\mathbb{F}_q$. $a, b \in \mathbb{F}_q$ are filed elements to determine equations of curve E. n is the order of a finite point $G = (x_G, y_G)$. Co-factor h is defined as $h = \#E(\mathbb{F}_q)/n$ where $\#E(\mathbb{F}_q)$ denotes the number of $\mathbb{F}_q$-points on E.

- ECDSA-KeyGen((pk, sk) $\xleftarrow{\$}$ param$_{ECDSA}$): On input system parameters param$_{ECDSA}$ = $\{q, FR, a, b, G, n, h\}$, this randomized algorithm outputs a private and public key pair (sk, pk). Here, for a random $1 \leq d \leq n-1$, return $Q = dG$. Set $sk = d$ and $pk = Q$.

- ECDSA-Sign(σ $\xleftarrow{\$}$ (param$_{ECDSA}$, sk, m)): On input system parameters param$_{ECDSA}$ = $\{q, FR, a, b, G, n, h\}$, user's private key sk and a message $m \in \{0, 1\}^*$, output signature $\sigma(m)$.

- ECDSA-Ver((0 or 1) $\leftarrow$ (pk, σ)): On input a signature $\sigma(m)$ and user's public key pk, verify the validity of signature $\sigma(m)$. Output 0 to signify an invalid signature; 1, otherwise.

4.3.2   Analysis of ECDSA's Security and Efficiency

Remark 4.1 For informal analysis, ECDSA signatures [103] are formed of pairs(r, s), where a random integer k is used to compute r. k is also used to compute s. We are able to calculate the private key if k is leaked. For example, leakage occurs due to low entropy sources of randomness. To improve this vulnerability, RFC6979 is suggested to strengthen the sources of randomness [104]. Meanwhile, reusing k and d also lead to revelation of private key, resulting in significant losses in Bitcoin history [104]. The possible attacks on ECDSA are: (1). Attacks on the intractability of ECDLP (2). Attacks on the underlying hash function of elliptic curve discrete logarithm problem (3). Other attacks, such as Vaudenay's attacks [105], Duplicate-signature key selection, implementation attacks, etc [103].

Tip 4.1 For formal analysis, the existentially unforgeable against a chosen message attack (EU-CMA) model we previously formalized in Section 4.2.3 need to be followed. Brown [106] proved the security of ECDSA under the assumption that the underlying hash function is collision-resistant and an underlying group is a generic group. In addition, some variants of DSA and ECDSA have also been proven to be secure in EU-CMA in ref. [107,108].

Next, we give Definition 4.2 as a precondition for the security analysis. Then, we focus on briefing the methodology to reduce attacks on ECDSA to the attacks against the preimage resistance or collision resistance of underlying hash function. Since the detailed analysis of ECDSA is proceeded mainly by mathematical illation, it does not involve tricky or skilled use of proof techniques. We omit details here and refer readers to ref. [106] for more details.

Theorem 4.1 *ECDSA scheme is existentially unforgeable against a chosen message attack if the underlying hash function H is collision-resistant and the underlying group is generic.*

Definition 4.2 *A (cryptographic) hash function H is a function that maps bit strings of arbitrary lengths to bit strings of a fixed length t such that:*

- **Computable.** *H can be computed efficiently;*

- **Preimage Resistance.** *For essentially all* $y \in \{0,1\}^t$ *it is computationally infeasible to find a bit string* x *such that* $H(x) = y$;

- **Collision Resistance.** *It is computationally infeasible to find distinct bit strings* x_1 *and* x_2 *such that* $H(x_1) = H(x_2)$.

Sketching the proof of Theorem 4.1. Suppose $\mathcal{A}$ is a probabilistic polynomial time adversary against ECDSA scheme. Then, the following explains how $\mathcal{A}$'s attacks on ECDSA can be successfully launched against the underlying hash function (say, SHA-1) which is not preimage resistant or collision resistant.

1. If SHA-1 is not preimage resistant, then an adversary $\mathcal{A}$ can forge signatures as follows. $\mathcal{A}$ selects an arbitrary integer l, and computes r as the x-coordinate of $Q + lG$ reduced modulo n. $\mathcal{A}$ sets $s = r$ and computes $e = rl \mod n$. If $\mathcal{A}$ can find a message m such that $e = H(m)$, then (r, s) is a valid signature for m.

2. If SHA-1 is not collision resistant, then $\mathcal{A}$ can repudiate signatures as follows. $\mathcal{A}$ first generates two messages m and m' such that SHA-1(m) = SHA-1(m') (known as a *collision*). $\mathcal{A}$ then signs m, and later claims to have signed m' (note that every signature for m is also a signature for m').

4.4 Case Analysis: BLS

This section reviews the construction, security requirement, and analysis of Boneh, Lynn and Shacham's (BLS) signature scheme [31].

4.4.1 Construction of BLS

We give the sketch of basic BLS construction in Figure 4.2. It makes use of a full-domain hash function $H : \{0,1\}^* \to G_1$ [22], basic construction and some other techniques to achieve short signatures [31].

Let (G_1, G_2) be Diffie-Hellman pair where $|G_1| = |G_2| = p$. A signature σ is an element of G_1, and comprises following the three algorithms.

- KeyGen. Choose random $x \overset{R}{\leftarrow} Z_p$ and compute $v \leftarrow g_2^x$. The public key is $v \in G_2$. The secret key is x.

- Sign Choose a secret key $x \in Z_p$, and a message $M \in \{0,1\}^*$, compute $h \leftarrow H(M) \in G_1$, and $\sigma \leftarrow h^x$. The signature is $\sigma \in G_1$.

Boneh, Lynn and Shacam's (BLS) Signature

System parameter: $\mathsf{param}_{\mathsf{BLS}} = \{G_1, G_2, H, p\}$.   Public key: x;Private key $v = g_2^x$

Signer	Verifier
1. Pick a message $M \in \{0,1\}^*$. 2. Compute $h \leftarrow H(M) \in G_1$. 3. Compute $\sigma \leftarrow h^x$. 4. Output signature $\sigma \in G_1$.	5. Given a signature $\sigma \in G_1$. 6. Compute $h \leftarrow H(M) \in G_1$. 7. Verify whether (g_2, v, h, σ) is a DDH tuple. 8. Output 1 if yes; output 0 if no.

FIGURE 4.2: Sketch of BLS Scheme

- Verification. Given a public key $v \in G_2$, a message $M \in \{0,1\}^*$, and a signature $\sigma \in G_1$, compute $h \leftarrow H(M) \in G_1$ and verify whether (g_2, v, h, σ) is a DDH tuple (as discussed in Definition 2.11 in Section 2). If yes, output 1; otherwise, output 0.

4.4.2   Security Requirement of BLS

Next, we define a game between a challenger, and an adversary $\mathcal{A}$ to capture existential unforgeability under a chosen message attack (UF-CMA) [4] for BLS scheme.

- **Setup.** The challenger runs algorithm KeyGen to output a public key PK and a private key SK. $\mathcal{A}$ is given the PK.

- **Queries.** $\mathcal{A}$ requests signatures with PK on at most q_s messages adaptively of his choice $M_1, \cdots, M_{q_s} \in \{0,1\}^*$. Denote each response to query as : $\sigma_i = \mathsf{Sign}(SK, M_i)$.

- **Output.** Finally, $\mathcal{A}$ outputs a pair (M^*, σ^*) and wins the game if (1) M^* is never queried, and (2) $\mathsf{Verify}(PK, M^*, \sigma^*) = 1$.

4.4.3   Security Analysis of BLS

Remark 4.2 As a high level, the security proof analysis of BLS [31] manages to reduce an attack on this scheme to solving an underlying hard problem. The associated proving techniques are proof by sequences of games, proof by theorem and case analysis. In the following, we will highlight complementary proving skills in the box and leave notes with the explanation. We mainly give three tips to help readers comprehend and study the proof skills step by step.

Tip 4.2 Following a game between a challenger and an adversary $\mathcal{A}$ specified in Section 4.4.2, we next give security analysis based on this model. To start, Boneh et al. [31] first defines the term (t, q_s, q_H, ϵ) for adversary $\mathcal{A}$ and existentially unforgeability in Definition 4.3. It also defines the premise of achieving existentially unforgeability under (t, q_s, q_H, ϵ). Noticeably, it reminds us to begin reduction with the assumption of a (t, q_s, q_H, ϵ)-adversary $\mathcal{A}$. Next, Boneh et al. [31] announce all critical preconditions and terms for achieving (t, q_s, q_H, ϵ)-existentially unforgeability in Theorem 4.2. This can seem as a non-self-evident statement which is about to be proven to be correct (by proof of Theorem 4.2). For ease of understanding, we remind the reader that the term "Theorem" is a form often used to present security analysis with the tidy format and mathematical manner. Here, we avoid further clarification on the formal use of terms like Definition, Theorem, Lemma, etc. The proof of Theorem 4.2 follows transitions based on failure events 2.4.2.2. More specifically, we remind the reader to notice the definition of **H-quries** and **Signature queries**. Although it might be confusing to comprehend the design philosophy initially, the reader should think more about it later after finishing the proof. Let's see details as below.

We define $\mathsf{AdvSig}_{\mathcal{A}}$ to be the probability that $\mathcal{A}$ wins in the above game, taken over the coin tosses of KeyGen and of $\mathcal{A}$.

Definition 4.3 *An adversary $\mathcal{A}(t, q_s, q_H, \epsilon)$ breaks a signature scheme if $\mathcal{A}$ runs in time at most t; $\mathcal{A}$ makes at most q_s signature queries and at most q_H queries to the hash function; and $\mathsf{AdvSig}_{\mathcal{A}}$ is at least ϵ. BLS scheme is (t, q_s, q_H, ϵ) existentially unforgeable under an adaptive chosen-message attack if no adversary (t, q_s, q_H, ϵ) breaks it.*

Theorem 4.2 *Let (G_1, G_2) be a (t', ϵ') pair of GDH group of order p. Then the signature scheme on (G_1, G_2) is (t, q_s, q_H, ϵ)-secure against existential forgery under an adaptive chosen-message attack (in the random oracle model) for all t and ϵ satisfying*

$$\epsilon \geq e(q_s + 1) \cdot \epsilon' \quad and \quad t \leq t' - c_{G_1}(q_h + 2q_s), \tag{4.1}$$

and c_{G_1} is a constant that depends on G_1. Here, e is the base of the natural logarithm.

Proof of Theorem 4.2. Suppose $\mathcal{A}$ is an adversary who (t, q_s, q_H, ϵ)-breaks the signature scheme. We show how to construct a t'-time algorithm $\mathcal{B}$ that solves CDHP in (G_1, G_2) with probability at least ϵ'.

Algorithm $\mathcal{B}$ is given $g_2, u \in G_2$ and $h \in G_1$, where $u = g_2^a$. The goal is to output $h^a \in G_1$. $\mathcal{B}$ simulates the challenger and interacts with the adversary $\mathcal{A}$ as follows.

- **Setup.** $\mathcal{B}$ starts by giving g_2 and $u \cdot g_2^r \in G_2$ to $\mathcal{A}$. Here, r is random in Z_p.

- **H-queries.** $\mathcal{A}$ can query the random oracle H at any time. $\mathcal{B}$ maintains query and response tuples (M_j, w_j, b_j, c_j) in the H-list. To respond, $\mathcal{A}$ proceeds as follows:

  1. If the query M_i already appears in the H-list in a tuple (M_j, w_j, b_j, c_j), $\mathcal{B}$ responds with $H(M_i) = w_i \in G_1$.
  2. Otherwise, $\mathcal{B}$ picks a random $c_i \in Z_p$ such that $\Pr[c_i = 0] = 1/(q_s + 1)$.
  3. $\mathcal{B}$ picks a random $b_i \in Z_p$. If $c_i = 0$, $\mathcal{B}$ computes $w_i \leftarrow h \cdot \psi(g_2)^{b_i} \in G_1$. If $c_i = 1$, $\mathcal{A}$ computes $w_i \leftarrow \psi(g_2)^{b_i} \in G_1$.
  4. $\mathcal{B}$ adds the tuple (M_j, w_j, b_j, c_j) to the H-list and responds to $\mathcal{A}$ by setting $H(M_i) = w_i$.

- **Signature queries.** Algorithm $\mathcal{B}$ responds to each signature query M_i as follow:

  1. $\mathcal{B}$ first runs corresponding H-queries to obtain a $w_i \in G_1$ such that $H(M_i) = w_i$. Let (M_i, w_i, b_i, c_i) be the corresponding tuple on the H-list. If $c_i = 0$ then $\mathcal{B}$ reports failure and terminates.
  2. If $c_i = 1$, set $w_i = \psi(g_2)^{b_i} \in G_1$. Then, $\mathcal{B}$ gives σ_i to algorithm $\mathcal{A}$ where $\sigma_i = \psi(u)^{b_i} \cdot \psi(g_2)^{rb_i} \in G_1$.

- **Output.** Finally, $\mathcal{A}$ produces a message-signature pair (M_f, σ_f) such that no signature query was issued for M_f. Assume σ_f is a valid signature on M_f under the given public key, $\mathcal{B}$ finds the tuple (M_f, w, b, c) on the H-list. If $c = 1$ then $\mathcal{B}$ reports failure and terminates. Otherwise, if $c = 0$, $\mathcal{B}$ outputs h^a as $h^a \leftarrow \sigma/(h^r \cdot \psi(u)^b \cdot \psi(g_2)^{rb})$.

Tip 4.3 To generalize, the above shows a construction of algorithm $\mathcal{A}$ which on input a CDHP instance $g_2, u = g_2^a$, it seeks to output $h^a \in G_1$. This is an example of proof by theorem. In short, theorem states non-self-evident statement which the author tries to prove it later.

First of all, the reader may ask: why do we define $c_i \in Z_p$ such that $\Pr[c_i = 0] = 1/(q_s + 1)$? Answer: it is helpful to just temporarily set $\Pr[c_i = 0] = \beta$ and determine β later.

As the reader should be aware of, there exist abortions during **H-queries** and **Signature queries** stages. This is an example of proof by transitions based on failure events. The failure creates events for use to analyze $\mathcal{A}$'s and $\mathcal{B}$'s probability in winning. First of all, the constructed $\mathcal{A}$ may not necessarily return a valid answer h to each given instance g_2, u. Nevertheless, does that mean $\mathcal{A}$ fail in the above game? The answer is No. According to the definition of probabilistic algorithm and security reduction given in Section 2.4, we here only seek to measure the probability of $\mathcal{A}$ in winning a game. In other words, the proposed game will be executed several times in order to measure the security.

Moreover, if adversary $\mathcal{A}$ always loses the game, it means that the query stages do not provide useful information for his attack. Conversely, if $\mathcal{A}$ manages to win the game by either accident or plotted attack, there must be ways to distinguish between the random guessing and the attack based on exploiting the previously launched queries. So, by case analysis, it allows us to determine $\mathcal{A}$'s winning probability and thus determine β where $\Pr[c_i = 0] = \beta$.

Knowing this, it remains to discuss the conditions and probabilities for abortions, and specifically focus on the successful case that yields a valid answer.

The above completes the description of the algorithm $\mathcal{A}$. It remains to show that $\mathcal{B}$ solves the given instance of the CDHP in (G_1, G_2) with probability at least ϵ'. To do so, we analyze the three events needed for $\mathcal{B}$ to succeed.

ε_1 : $\mathcal{B}$ does not abort as a result of any of $\mathcal{A}$'s signature queries.

ε_2 : $\mathcal{A}$ generates a valid message-signature forgery (M_f, σ_f).

ε_3 : Event ε_2 and $c = 0$ for the tuple containing M_f on the H-list.

$\mathcal{B}$ succeeds if all of these events happen. The probability $\Pr[\varepsilon_1 \wedge \varepsilon_3]$ decomposes as:

$$\Pr[\varepsilon_1 \wedge \varepsilon_3] = \Pr[\varepsilon_1] \cdot \Pr[\varepsilon_2|\varepsilon_1] \cdot \Pr[\varepsilon_3|\varepsilon_1 \wedge \varepsilon_2]. \tag{4.2}$$

The following claims give a lower bound for each of these terms.

Claim 1: The probability that algorithm $\mathcal{B}$ does not abort as a result of $\mathcal{A}$'s signature queries is at least $1/e$. Hence, $\Pr[\varepsilon_1] \geq 1/e$.

Claim 2: If algorithm $\mathcal{B}$ does not abort as a result of $\mathcal{A}$'s signature queries then algorithm $\mathcal{A}$'s view is identical to its view in the real attack. Hence, $\Pr[\varepsilon_2|\varepsilon_1] \geq \epsilon$.

Claim 3: The probability that algorithm $\mathcal{B}$ does not abort after $\mathcal{A}$ outputs a valid forgery is at least $1/(q_s + 1)$. Hence, $\Pr[\varepsilon_3|\varepsilon_1 \wedge \varepsilon_2] \geq 1/(q_s + 1)$.

Proof of Claim 1, 2 and 3: Can be found in ref. [31] on page 6. We leave the proofs of these claims to the reader to complete.

Using the bounds from the claims above, one can prove that $\mathcal{B}$ produces the correct answer with probability as least $\epsilon/e(q_s + 1) \geq \epsilon'$ as required. Following this, it is easy to complete the proof of Theorem 4.2.

Tip 4.4 To determine $\mathcal{A}$'s winning probability and $\mathcal{B}$'s abortion probability, three events as necessities for $\mathcal{B}$ to output valid answer are analyzed: $\varepsilon_1, \varepsilon_2, \varepsilon_3$. To note, event ε_2 is a precondition for event ε_3. However, event ε_1 is independent of ε_2. Therefore, it needs to define $\varepsilon_1, \varepsilon_2, \varepsilon_3$ separately for discussion. Consequently, $\mathcal{B}$'s winning case can be expressed by: $\varepsilon_1 \wedge \varepsilon_3$. Accordingly, this equation can be further decomposed into three independent cases $\varepsilon_3 | \varepsilon_1 \wedge \varepsilon_2$ for the analysis. To do the case analysis for each condition, the corresponding claim is given. Typically, case analysis is mainly used in this part as the main proving technique for the analysis. Since we find the proof of claim easy to follow at this point, we suggest readers to do it independently.

4.5   Group Signature and Case Analysis

In this section, we review the group signature and instantiate it with a case scheme called MLSAG. We then analyze the construction and security of MLSAG.

4.5.1   Background of Group Signature

The development of group signature can be summarized as the following stages.

Stage one (1991–1997): Chaum and Vail Heyst [75] gave the definition and four concrete constructions for group signature. In 1995, Chen and Pedersen [109][proposed two group signatures to address open problems issued by ref. [75]. After several years, cryptographic researchers started to focus on the group signature. During this stage, Park et al. [110] proposed the first identity-based group signature. Mao and Lim [111] pointed out that Park et al.'s [110] did not achieve anonymity. Then, Tseng and Jan proposed a new identity-based group signature [112] without satisfying unforgeability and collusion resistance [113,114]. The proposed schemes during this stage had low efficiency and linear relationship between the group public key and the number of group members.

Stage Two (1997–2000): Camenisch and Stadler [115] proposed a group signature ideal for large groups. In this proposal, the group public key's size was independent of the number of group members. Meanwhile, the signing and verification algorithms achieved complexity, which is independent of the group size. After that, researches of this field entered to an active era. During that time, researches focused more on the practicality and efficiency of group signatures. To note, Camenisch and Michels proposed efficient group signature schemes [116,117]. Conversely, Kim et al. [118] showed how to transform a group signature scheme into an ordinary signing scheme. Castelluccio [119] discussed how to transform identity-based signing scheme into an ordinary digital signing scheme. In 1997, Kim et al. [120] proposed hierarchical group signature which allowed a higher rank member to sign on behalf of the whole group and represented lower rank member to sign without revealing higher rank members' certificates. Later, Lysyanskaya [121] proposed a blind signature notion and pointed out how to use it to construct anonymous digital currencies co-issued by multiple banks. In 1999, Ateniese and Tsudlk [122] proposed multi-group and subgroup signing schemes. This allowed a signer belonged to multiple groups to sign on behalf of these groups.

Stage Three (2003–2007): The use of pairing computation in group signature produced short signature length and application-rich features for the group signature. Chen [123] proposed the first bilinear pairing and identity-based group signature where a private key generator is employed to solve the escrow problem. Wei et al. [124] tried to re-define the notion of identity-based group signature by proposing a new and complete identity-based group signature scheme where the authorities of group manager and member in opening a signature are identity-based. Also, the notion of threshold group signature and standard model-based schemes were proposed.

Stage Four (2008–present): Infrastructures like blockchain, AI, Internet-of-Things (IoT) have driven the exhaustive researches and diverse applications of group signature. In 2009, Nakanishi et al. [125] proposed a group signature with constant complexity in signing and the verification. In 2011, Fan et al. [126] proposed accumulator based group signature with the group membership revocation, which required a trusted private key generator to assist group manager with group member's revocation. Meanwhile, the revocation required $\sqrt{n}$ computing complexity for each non-revoked group member. In 2012, Libert et al. [127] proposed the group signature with a novel revocation technique for large scales of group members in the standard model $\mathcal{O}(1)$ verification. Ring signature (discussed in Section 4.6) is another important primitive transformed from the group signature. These two notions have been used as basic building blocks for the blockchain projects, like Monero [12,102,128].

4.5.2 Definition of Group Signature

Following Bellare et al.'s work [129], we give formal definitions of group signature as follows.

A group signature scheme $\mathcal{GS} = (\mathsf{GKeyGen}, \mathsf{GSign}, \mathsf{GVer}, \mathsf{Open})$ consists of four polynomial-time algorithms:

- $\mathsf{GKeyGen}((gpk, gmsk, gsk) \xleftarrow{\$} \lambda)$: On input a security parameter λ, it outputs group public key gpk, group manager's secret key $gmsk$ and an n-vector of keys consists of n secret signing key $gsk[i]$ for the group member with gsk such that $gsk = \{gsk[i]\}_{1 \le i \le n}$.

- GSign($\sigma \xleftarrow{\$} (gsk[i], m)$): On input a secret signing key $gsk[i]$ and a message with arbitrary length $m \in \{0,1\}^*$, it outputs a signature σ.

- GVer((0 or 1) $\leftarrow (gpk, m, \sigma)$): On input gpk, m and σ, it outputs 0 or 1.

- Open((id or $\perp$) $\leftarrow (gmsk, m, \sigma)$): On input $gmsk$, m and σ, it outputs an identity id of the actual signer or the symbol $\perp$ to indicate failure.

4.5.3   Case Study: MLSAG Scheme

4.5.3.1   Construction of MLSAG

Multi-layered Linkable Spontaneous Anonymous Group Signature (MLSAG)

System parameter: $\text{param}_{\text{MLSAG}} = \{G, p, H_p, l\}$

Public key: $P_j = xG$; Private key: x. Denote π as the index of actual signer

Signer	Verifier
1. For $j = 1, \cdots, m$, set $I_j =_j H(P_\pi^j)$ and pick $\alpha_j \in Z_q$. Compute $L_\pi^j = \alpha_j G$ and $R_\pi^j = \alpha_j H(P_\pi^j)$.	7. Given a signature σ, for $i = 1, \cdots, n$, compute each L_i^j, R_i^j and c_i.
2. For $j = 1, \cdots, m, i = 1, \cdots, \hat{\pi}, \cdots, n$, set $s_i^j \xleftarrow{R} Z_q$.	8. Compute $c_{n+1} = H(m, L_n^1, R_n^1, \cdots, L_n^m, R_n^m)$
3. For $j = 1, \cdots, m$, compute $c_{\pi+1} = H(m, L_\pi^1, R_\pi^1, \cdots, L_\pi^m, R_\pi^m)$, $L_{\pi+1}^j = s_{\pi+1}^j G + c_{\pi+1} P_{\pi+1}^j$, $R_{\pi+1}^j = s_{\pi+1}^j H(P_{\pi+1}^j) + c_{\pi+1} I_j$.	9. Check whether $c_{n+1} \overset{?}{=} c_1$.
4. Then, for $i = 1, \cdots, n$, compute $L_{\pi-1}^j = s_{\pi-1}^j G + c_{i-1} P_{i-1}^j$, $R_{\pi-1}^j = s_{\pi-1}^j H(P_{i-1}^j) + c_{i-1} \cdot I_j$.	10. If yes, output 1; otherwise, output 0.
5. For $j = 1, \cdots, m$, compute s_π^j from $\alpha_j = s_\pi^j + c_\pi x_j$ mod ℓ.	
6. Derive signature $\sigma = (I_1, \cdots, I_m, c_1, s_1^1, \cdots, s_1^m, s_2^1, \cdots, s_2^m, \cdots, s_n^1, \cdots, s_n^m)$.	

FIGURE 4.3: Sketch of MLSAG Scheme

We give a sketch of multilayered linkable spontaneous anonymous group signature (MLSAG) proposed by Noether and Mackenzie in ref. [102] in Figure 4.5. As we earlier introduced in Section 4.2.2.2, this proposal seeks to achieve hidden amounts of any transactions in Monero [44] Detailed construction are as follows.

Assume each signer of a ring consists of n members has m keys $\{P_i^j\}_{j=1,\cdots,m}^{i=1,\cdots,n}$. Given a message m to be signed, let π be a secret index corresponding to the actual signer. For $j = 1, \cdots, m$, let $I_j = x_j H(P_\pi^j)$, and for $j = 1, \cdots, m, i = 1, \cdots, \hat{\pi}, \cdots, n$, let s_i^j be some random scalars. Set:

$$L_\pi^j = \alpha_j G, \tag{4.3}$$

$$R_\pi^j = \alpha_j H(P_\pi^j). \tag{4.4}$$

For random scalars α_j and $j = 1, \cdots, m$. Next, compute:

$$c_{\pi+1} = H(m, L_\pi^1, R_\pi^1, \cdots, L_\pi^m, R_\pi^m), \tag{4.5}$$

$$L_{\pi+1}^j = s_{\pi+1}^j G + c_{\pi+1} P_{\pi+1}^j, \tag{4.6}$$

$$R_{\pi+1}^j = s_{\pi+1}^j H(P_{\pi+1}^j) + c_{\pi+1} I_j, \tag{4.7}$$

and repeat this, incrementing i modulo n until deriving:

$$L_{\pi-1}^j = s_{i-1}^j G + c_{i-1} P_{i-1}^j, \tag{4.8}$$

$$R_{\pi-1}^j = s_{i-1}^j H(P_{i-1}^j) + c_{i-1} \cdot I_j, \tag{4.9}$$

$$c_\pi = H(m, L_{\pi-1}^1, R_{\pi-1}^1, \cdots, L_{\pi-1}^m, R_{\pi-1}^m). \tag{4.10}$$

Finally, we solve each s_π^j using $\alpha_j = s_\pi^j + c_\pi x_j \mod \ell$. The signature is then given as $\sigma(m) = (I_1, \cdots, I_m, c_1, s_1^1, \cdots, s_1^m, s_2^1, \cdots, s_2^m, \cdots, s_n^1, \cdots, s_n^m)$. the Verification proceeds by regenerating all the L_i^j, R_i^j from $i = 1$ and checking whether $c_{n+1} = c_1$.

4.5.3.2 Security Analysis of MLSAG

Proof Overview 4.1 In a nutshell, based on previous work of Liu et al. [130], Shen and Mackenzie [102] can reduce the unforgeability of their proposed MLSAG to previous work (Linkable Spontaneous Anonymous Group Signature). By defining a security model for unforgeability and following the lemma and thr theorem which have been proved in previous work, Shen and Mackenzie give the security analysis of MLSAG.

They first define the preconditions for achieving unforgeability of MLSAG in Definition 4.4, where the forged signature produced by adversary $\mathcal{A}$ is supposed to be unknown to $\mathcal{A}$ and hidden among a list of keys $\mathcal{L}$ (as a vector). The proof techniques used in this scheme are: the proof by contradiction, and the rewind simulation [107,130,131]. A sketch of proof is given as follows.

Definition 4.4 Unforgeability. *A MLSAG signature scheme is unforgeable if for any probabilistic polynomial time (PPT) algorithm $\mathcal{A}$, given a list of n public key vectors chosen by $\mathcal{A}$, then $\mathcal{A}$ can only produce a valid signature with negligible probability when $\mathcal{A}$ does not know one of the corresponding private key vectors.*

Tip 4.5 Next, Shen and Mackenzie create a probabilistic polynomial time adversary $\mathcal{M}$ which uses $\mathcal{A}$ to find the discrete logarithm of one of the keys in $\mathcal{L}$. This part's proving technique is by contradiction.

Proof of Unforgeability. This follows as [130], Theorem 1. Let H_1 and H_2 be random oracles, and $\mathcal{SO}$ be a signing oracle which returns valid MLSAG signatures. Assume $\mathcal{A}$ as an adversary who efficiently breaks MLSAG scheme, i.e. the ability to forge an MLSAG signature from a list of key vectors L.

$$\Pr(\mathcal{A}(\mathcal{L}) \rightarrow: \mathrm{Ver}(L, m, \sigma) = \mathrm{True}) > \frac{1}{Q_1 k}. \tag{4.11}$$

Assume that $\mathcal{A}$ makes no more than $q_H + nq_s$ queries to the signing oracles H_1, H_2 and $\mathcal{SO}$. We next show how to construct a probabilistic polynomial time (PPT) $\mathcal{M}$ which uses $\mathcal{A}$ to solve DLP of one of the keys in $\mathcal{L}$. If $\mathcal{L}$ is a set of key vectors $\{\overline{y_1}, \cdots, \overline{y_n}\}$ each of size r, then a forged signature $\sigma = (c_1, s_1, \cdots, s_n, y_0)$ must satisfy:

$$c_{i+1} = H(m, L - i^1, R_i^1, \cdots, L_i^m, R_i^m). \tag{4.12}$$

$\mathcal{M}$ may call $\mathcal{A}$ to forge signatures several times and record each script history in a list.

Tip 4.6 Next, Shen and Mackenzie intend to prove Lemma 4.1. The reason is that Lemma 4.1 is useful to validate the proof of Theorem 4.3, and Theorem 4.3 state that the attack from adversary $\mathcal{A}$ can be reduced to solve the DLP. This forms a desired security reduction.

The proof of Lemma 4.1 is given in an original work of Liu et al. [130] by forking lemma [107]. It defines a lower bound for the success probability of $\mathcal{A}$ with given transcript header H and $\mathcal{M}$. Here, transcript header H is a notion in the rewind simulation which relates a simulation transcript tape $\mathcal{T}$ used to record the coin flip sequences. Specifically, following Liu et al. 's work [130], Theorem 4.3 can be proved by using a very simple adaptive mechanism called rewind on success.

In the end, we would like the reader to refer to ref. [130] for more details. To add, it is acceptable to prove one scheme's security by following theorem or lemma addressed by another work. However, we should know that this can only work if: (1) Two schemes are a similar design (2) The underlying security model and associated definitions are properly defined; Otherwise, it seems like a false claim or an indiscreet move.

Lemma 4.1 *[Lemma 1 [130]] Let $\mathcal{M}$ invoke $\mathcal{A}$ to obtain a transcript $\mathcal{T}$: If $\mathcal{T}$ is successful, then $\mathcal{M}$ rewinds $\mathcal{T}$ to a header H and re-simulates $\mathcal{A}$ to obtain transcript $\mathcal{T}'$. If $Pr(\mathcal{T}\,succeeds) = \epsilon$, then $Pr(\mathcal{T}'\,succeeds) = \epsilon$.*

Proof. Follows from the proof of Lemma 1 in ref. [130].

Theorem 4.3 *The probability that an adversary $\mathcal{A}$ forges a valid MLSAG signature is negligible under DLP.*

Proof of Theorem 4.3. Similar to the proof of Theorem 1 given in ref. [130], in that the probability of guessing the output of a random oracle is negligible. Then, there are $\mathcal{T}$ queries to H_1 matching the n queries used to verify the signature. Let $X_{i_1}, \cdots, X_{i_m}$ denote queries used in the verification for the ith such forgery and let π be the index of an actual signer, then we have:

$$X_{I_M} = H_1(m, L_{\pi-1}^1, R_{\pi-1^1}, \cdots, L_{\pi-1}^{m\mathcal{T}}, r_{\pi-1}^{m\mathcal{T}}). \tag{4.13}$$

An attempted forgery σ produced by $\mathcal{A}$ is an (ℓ, π)-forgery if $i_1 = \ell$ and π is as above. By assumption, there exists a pair (ℓ, π) such that the probability that the corresponding transcript $\mathcal{T}$ gives a successful forgery $\epsilon_{\ell,\pi}(\mathcal{T})$, satisfies

$$\epsilon_{\ell,\pi} \geq \frac{1}{m_\mathcal{T}(q_H + m_\mathcal{T}q_s)}) \cdot \frac{1}{Q_1(k)} \geq \frac{1}{n(q_H + nq_S)} \cdot \frac{1}{Q_1(k)}. \tag{4.14}$$

Now, rewinding $\mathcal{T}$ to just before the ℓth query, it follows that the probability that $\mathcal{T}'$ is also a successful forgery satisfies

$$\epsilon_{\ell,\pi}(\mathcal{T}') \geq \frac{1}{n(q_H + nq_s)} \cdot \frac{1}{Q_1(k)}. \tag{4.15}$$

Intuitively, the probability that both $\mathcal{T}$ and $\mathcal{T}'$ correspond to verifying forgeries σ and σ' is non-negligible:

$$\epsilon_{l,\pi}(\mathcal{T} \text{ and } \mathcal{T}') \geq (\epsilon_{l,\pi}(\mathcal{T}))^2. \tag{4.16}$$

New coin-flips have been computed, and it follows that with overwhelming probability there is j such that $s_\pi^j \neq s_\pi'^j$ and $c_\pi \neq c_{\pi+1}$. Thus, we can compute the private key of index π by:

$$x_\pi^j = \frac{s_\pi'^j - s_\pi^j}{c_\pi - c_\pi'} \mod q. \tag{4.17}$$

This completes the proof.

4.5.4   Efficiency Analysis of MLSAG

For ease of comparison, we borrow results from a peer work: RingCT3.0 [128] to evaluate the efficiency of MLSAG (RingCT1.0). To explain, Yuen et al. [128] proposed an improved blockchain ring confidential transaction protocol (RingCT3.0) from previous works RingCT2.0 [11] and RingCT1.0 [102]. Here, RingCT1.0 protocol is driven by MLSAG. As given in Figure 4.4, the performance of Spend (mainly decided by signing of MLSAG) and Verify (mainly decided by verifying MLSAG) of RingCT1.0 are compared with other protocols for the evaluation. Specifically, for Spend phase of the protocol, RingCT3.0 outperforms RingCT1.0 in large rings and larger user inputs. Also, in the case of the Verify protocol, RingCT3.0 outperforms RingCT1.0 as well. In general, the running time of RingCT3.0 is comparatively shorter than generating a block of transactions, which is 2 minutes in Monero and 10 minutes in Bitcoin. Therefore, RingCT3.0 is not the bottleneck of the blockchain system. For complexity analysis of MLSAG, since MLSAG is similar in design with LRRS (the next case), we refer readers to Section 4.6.2.4 for details.

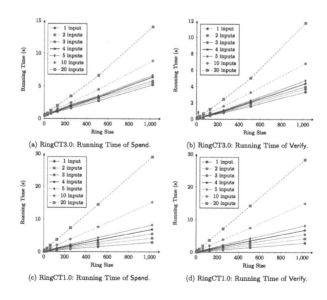

(a) RingCT3.0: Running Time of Spend.

(b) RingCT3.0: Running Time of Verify.

(c) RingCT1.0: Running Time of Spend.

(d) RingCT1.0: Running Time of Verify.

FIGURE 4.4: The Performance Comparison of RingCT1.0 and RingCT3.0

4.6 Ring Signature and Case Analysis

In this section, we review the notion of ring signature. For case study, we review the construction, security requirement and analysis, and complexity of our original proposal, called: linkable and redactable ring signature [132].

4.6.1 Background of Ring Signature

In 2001, Rivest, Shamir, and Tauman [76] first proposed the ring signature notion. Ring signature is characterized by two main properties: anonymity and spontaneity. Ring signature originated from group signature [75]. Group signature allows a real signer's anonymity to be revoked by a trusted party called group manager. However, this centralized trust reliance contradicts with the decentralized feature of blockchain. Ring signature, on the other hand, does not allow anyone to revoke the signer anonymity while allowing the real signer to form a group (also known as a ring) arbitrarily without being controlled by any other party. In short words, a signer can randomly choose a group of users to form a ring and sign a signature on behalf of the whole group. Other group members are unaware of the fact that they have been summoned. So, the formation of the ring is spontaneous, and full anonymity is thus achieved. Dominant security and efficiency issues have been discussed in [31,115,133]. There are different types of ring signatures including threshold version [134,135], revocable version [136,137] and linkable version [12,77,138,139].

The development of crypto-currencies has driven the researches of ring signature. Recently, a linkable ring signature is gaining massive attention, as it allows detection of double-spending in cryptocurrencies. It also caters for anonymity and security requirements [1,140] simultaneously. To explain, it can link a double-signed signature to avoid any double-spending payment in blockchains [141]. We review some relevant works in Table 5.3. As it is shown in Table 5.3, most ring signature schemes are based on public-key infrastructure and are linkable. In addition, few works achieve the design of constant signature size (i.e. $|\sigma| = \mathcal{O}(1)$).

TABLE 4.1: Overview of Ring Signature Works

Scheme	Crypto-system	Pairing-based	Signature Size	Link-ability	Assump-tions
SLT [138]	PKI	No	$\mathcal{O}(n)$	Yes	DDHP, S-RSA
SLRSE [77]	PKI	No	$\mathcal{O}(1)$	Yes	S-RSA
SLRSR [142]	PKI	No	$\mathcal{O}(n)$	Yes	DDHP, S-RSA
LRSS [139]	PKI	No	$\mathcal{O}(n)$	Yes	DDHP
RCT2.0 [12]	PKI	Yes	$\mathcal{O}(1)$	Yes	DDH, k-SDH
LRSU [143]	PKI	No	$\mathcal{O}(n)$	Yes	DLP
ELT [144]	PKI	Yes	$\mathcal{O}(\sqrt{n})$	Yes	SDH, DDHI
SIB [145]	IDB	Yes	$\mathcal{O}(1)$	Yes	DLP, DDHP
ESL [146]	PKI	Yes	$\mathcal{O}(\sqrt{n})$	Yes	SDA, q-SDH
TRS [147]	PKI	No	$\mathcal{O}(n)$	No	DDHP

Denote PKI as Public-Key based Infrastructure; IDB as Identity-based Infrastructure.

4.6.2   Case Study: LRRS Scheme

4.6.2.1   Construction of LRRS

Liu et al. [130] proposed a variant of ring signature schemes. Inspired by their work and motivated by redactable blockchain (as will discuss in Section 8.5), we propose a linkable and redactable ring signature (LRRS) in ref. [132]. A sketch of the LRRS scheme is given in Figures 4.5 and 4.6. We only give a brief introduction to each algorithm and omit detailed construction due to space limitation. However, the provisions of constructional sketch and algorithm definitions should suffice for readers to understand its basic idea. We refer the reader to ref. [132] for more details.

Linkable and Redactable Ring Signature (LRRS)

System parameter: $\mathsf{param}_{\mathsf{LRRS}} = \{G, q, g, H_1, H_2, \Delta t\}$

Public key: $hk = y = g^x$ and $pk_i = y_i = g^{x_i}$; Private key: $tk = x$, $sk_i = x_i \in Z_q$.

Signer	**Verifier**
1. Compute $h = H_1(\mathsf{CID})$ and $\tilde{y} = h^{x_\pi}$.	8. For $i = 1, \cdots, n$, compute $\hbar_i = g^{\alpha_i t} h^{H_2(m)\bar{t}}$.
2. Pick $u, v \xleftarrow{R} Z_q$ and compute $c_{\pi+1} = H_1(L, \tilde{y}, g^{ut}, h^v)$.	9. Compute $z_i' = \hbar_i \cdot y_i^{c_i}$, $z_i'' = h^{\beta_i} \tilde{y}^{c_i}$.
3. For $i \neq \pi$, pick $\alpha_i, \beta_i \xleftarrow{R} Z_q^*$, compute $r_i = (g^{\alpha_i t}, y^{\alpha_i t})$ and $\hbar_i = g^{\alpha_i t} h^{H_2(m)t}$.	10. For $i \neq n$.
4. For $i \neq \pi$, compute $c_{i+1} = H_2(L, \tilde{y}, \hbar_i \cdot y_i^{c_i}, h^{\beta_i} \tilde{y}^{c_i})$.	Compute $c_{i+1} = H_2(L, \tilde{y}, z_i', z_i'')$
5. Compute $\alpha_\pi = u - x_\pi c_\pi t^{-1} \bmod q$.	
6. Compute $\beta_\pi = v - x_\pi c_\pi \bmod q$.	11. Check whether
7. Output signature $\sigma_L^t(m) =$ $(c_1, r_1, \cdots, r_n, \beta_1, \cdots, \beta_n, \tilde{y})$.	$c_1 \overset{?}{=} H_2(L, \tilde{y}, z_n', z_n'')$.
	If yes, output 1; otherwise, output 0.

FIGURE 4.5: Sketch of LRRS Scheme (a)

Linkable and Redactable Ring Signature (LRRS)
System parameter: $\text{param}_{\text{LRRS}} = \{G, q, g, H_1, H_2, \Delta t\}$
Public key: $hk = y = g^x$ and $pk_i = y_i = g^{x_i}$; Private key: $tk = x$, $sk_i = x_i \in Z_q$.

Redactor	Updater
1. Given $\sigma_L^t(m)$, first check its validity.	5. Given $\sigma_L^t(m)$, first check its validity.
2. If valid, proceed; otherwise, output $\bot$.	6. If valid, proceed; otherwise, output $\bot$.
3. Given m', for $1 \leq i \leq n$, compute $r_i' = (g^{\alpha_i' t} h^{((H_2(m)} {}_{-H_2(m'))t}, y^{\alpha_i' t} h^{x((H_2(m) - H_2(m'))t)})$.	7. Given $\Delta t = t' - t$, for $1 \leq i \leq n$, compute $r_i'' = (g^{\alpha_i^t} h^{H_2(m)(-\Delta t)}, y_i^{\alpha_i^t} h^{x H_2(m)(-\Delta t)})$.
4. Output a redacted signature $\sigma_L^t(m') = (c_1, r_1', \cdots, r_n', \beta_1, \cdots, \beta_n, \tilde{y})$.	8. Output an updated signature $\sigma_L^{t+\Delta t}(m) = (c_1, r_1'', \cdots, r_n'', \beta_1, \cdots, \beta_n, \tilde{y})$.

FIGURE 4.6: Sketch of LRRS Scheme (b)

- **LRRS.Setup**$(\lambda) \rightarrow (\text{param}_{\text{LRRS}})$. Initiate the parameters for whole system.

- **LRRS.UKeyGen**$(\text{param}_{\text{LRRS}}) \rightarrow (sk_i, pk_i)$. Generate private key and public key.

- **LRRS.HKeyGen**$(\text{param}_{\text{LRRS}}) \rightarrow (hk, tk)$. Generate trapdoor key (master key) and hash key.

- **LRRS.Sign**$(hk, L, m, x_\pi, t) \rightarrow (\sigma_L^t(m))$. Generate a LRRS signature. Here, denote L as a list of n public keys L, x_π as signer's private key, $\pi \in [1, |L|]$ as confidential and hidden.

- **LRRS.Verify**$(L, m, \sigma_L^t(m), \bar{t}) \rightarrow (0 \text{ or } 1)$. Verify validity of a given LRRS signature.

- **LRRS.Redact**$(tk, L, m', \sigma_L^t(m), t) \rightarrow (\bot \text{ or } \sigma_L^t(m'))$. Redact a given signature to re-commit to another new message without impairing verification.

- **LRRS.Update**$(tk, L, m, \sigma_L^t(m), t, \Delta t) \rightarrow (\bot \text{ or } \sigma_L^{t+\Delta t}(m))$. Update a given signature such that it can be verified at a given time.

- **LRRS.Link**$(L, (pk_{d_0}, m_{d_0}, \sigma_L^t(m_{d_0})), (pk_{d_1}, m_{d_1}, \sigma_L^t(m_{d_1})), \bar{t}) \rightarrow (\bot,$ $0 \text{ or } 1)$. Check whether two signature are from one identical signer.

- **LRRS.Deny**$(L, m^*, \sigma_t^L(m^*), m, \sigma_t^L(m), \bar{t}) \rightarrow (\bot, 0 \text{ or } 1)$. Solve a dispute on a given signature.

4.6.2.2   Security Requirement of LRRS

We capture two security requirements by defining follow security models.

Unforgeability: Our proposed LRRS is unforgeable against chosen-message attack (UNF-CMA) if for any PPT adversary $\mathcal{A}$, the probability in the following $UNF-CMA_{\mathcal{A}}^{\lambda}$ experiment is negligible $\Pr[UNF-CMA_{\mathcal{A}}^{\lambda} = 1] \leq \nu(\lambda)$ where ν is a negligible function.

Experiment: UNF-CMA$_{\mathsf{LRRS}}^{\mathcal{A}_1}(\lambda)$

    $\mathsf{param}_{\mathsf{LRRS}}^{ch} \leftarrow \mathsf{LRRS.Setup}(\lambda)$

    For L, n and t, where $\forall 1 \leq i \leq n$, $(sk_i, pk_i) \leftarrow$
    $\mathsf{LRRS.UKeyGen}(\mathsf{param}_{\mathsf{LRRS}}^{ch})$

    $(tk_{ch}, hk_{ch}) \leftarrow \mathsf{LRRS.UKeyGen}(\mathsf{param}_{\mathsf{LRRS}}^{ch})$

    $(m^*, \sigma_L^t(m^*)) \leftarrow \mathcal{A}^{\mathsf{Sign}(\cdots), \mathsf{Redact}(\cdots), \mathsf{Update}(\cdots)}(L, t)$

        where Sign, Redact and Update are signing,
        redacting and update oracles which will return $\perp$
        if queried on invalid inputs and response for
        at most q_s distinct queries in total

    Return 1 if $\mathsf{LRRS.Verify}\,(L^*, m^*, \sigma_{L^*}^t(m^*), \bar{t}) = 1 \wedge (L^* \subseteq L)$

        $\wedge \sigma_{L^*}^t(m^*) \neq \sigma_L^t(m^i)$ where $i \in [1, q_s]$

    and $\bar{t}$ is the current time

    ; otherwise, return 0.

Signer-Ambiguity: Our proposed LRRS is signer-ambiguous if the probability of adversary $\mathcal{A}$ in winning the following experiment is exactly $\frac{1}{|L|}$, i.e. $|\,Pr[SA_{\mathcal{A}}^{\lambda}]\,| = \frac{1}{|L|}$.

Experiment: $SA_{\mathcal{A}}^{\lambda}$

$\forall \mathsf{param}_{\mathsf{LRRS}} \leftarrow \mathsf{LRRS.Setup}(\lambda)$

    $\forall (tk, hk) \leftarrow \mathsf{LRRS.HKeyGen}(\mathsf{param}_{\mathsf{LRRS}})$

    Given $(L^*, n^*, \sigma_L^t(m^*))$ where $L^* = \{pk_1^*, \cdots, pk_n^*\}$
    and $\forall 1 \leq i \leq n^*$, $(sk_i^*, pk_i^*) \leftarrow \mathsf{LRRS.UKeyGen}\,(\mathsf{param}_{\mathsf{LRRS}})$

    $j \xleftarrow{\$} \{1, n\}$

    $j^* \leftarrow \mathcal{A}_1^{\mathsf{Sign}(\cdots), \mathsf{Redact}(\cdots), \mathsf{Update}(\cdots)}(L^*)$ where oracles
    Sign, Redact and Update are as previously defined

    Return 1 if $(j = j^*) \wedge (\forall i, j \in [1, n^*], \mathsf{LRRS.Link}() = 0)$

$\wedge\,(pk_j^*$ is not on the query list of oracles $\mathsf{Sign}, \mathsf{Redact}$ or $\mathsf{Update}))$

Otherwise, return 0

4.6.2.3 Security Analysis of LRRS

Proof Overview 4.2 Our LRRS scheme shares a similar design philosophy of Liu et al. [130]. However, we encapsulate the design of chameleon hash (trapdoor commitment) into the design of ring signature. As a result, it allows the signer to designate a trapdoor key holder. The generated ring signature can be redacted (recommitted to a new message) while still passing verification. the security of our LRRS mainly includes unforgeability and security of underlying chameleon hash (as given in Section 6.4.3).

Based on the security model of our LRRS, we can then prove its unforgeability by following proof techniques, like proof by contradiction, rewind simulation [107,130,131] as we previously mentioned. A sketch of proof is given as follows.

Proof of Unforgeability. reduce the existential unforgeability against chosen-message attack (UNF-CMA) of our LRRS scheme to the intractability of DLP in the random oracle.

Briefly, suppose $\mathcal{A}$ is a probabilistic polynomial time (PPT) adversary against the unforgeability of our LRRS. Suppose $\mathcal{A}$ makes at most q_s queries to Sign, Redact and Update oracles then $\mathcal{A}$ can forge a valid LRRS signature with non-negligible probability. Then, we can construct a simulator $\mathcal{M}$ which uses $\mathcal{A}$ to solve the DLP with one of the public keys in L. Denote the successful forgery output by $\mathcal{A}$ as below:

$$\sigma^* = (c_1^*, r_1^*, \cdots, r_n^*, \beta_1^*, \beta_n^*, \tilde{y}^*) \tag{4.18}$$

where the forged signature σ^* by $\mathcal{A}$ against our unforgeability experiment satisfies the following equations:

$$c_{i+1}^* = H_2(L, \tilde{y}, \hbar \cdot y_i^* c_i^*, h^{\beta_i^*} \tilde{y}^{c_i^*}), \tag{4.19}$$

$$c_1^* = H_2(L, \tilde{y}, \hbar \cdot y_n^{* c_n^*}, h^{\beta_n^*} \tilde{y}^{c_n^*}). \tag{4.20}$$

On input a DLP instance (g, g^x), $\mathcal{M}$ sets $y_{x_\pi} = g^x$ where $x_\pi \in L$, and it aims to output x. $\mathcal{M}$ records the queries history of $\mathcal{A}$ in a list and returns identical response upon receiving the same query. Some outputs of $\mathcal{A}$ are valid forgeries of LRRS and are used to solve DLP with non-negligible probability. The proof is similar to the one in Section 4.5.3.2.

Proof of Signer-Ambiguity. Our proposed LRRS is Signer-Ambiguous if the DDHP is hard in the random oracle. Briefly, suppose $\mathcal{A}$ is a PPT adversary against the signer-ambiguity of our LRRS, then we can construct a PPT simulator $\mathcal{M}$ to solve DDHP by using $\mathcal{A}$. Concretely, on given a message m, a list of n public keys L, a set of private keys $SKT = \{sk_i, \cdots, sk_t\}$ where $0 \leq t \leq n-1$, a signature σ_L^t signed with private key sk_π at current time t on message m, $\mathcal{M}$ can be constructed to solve DDHP for some polynomial $Q(k)$ with probability:

$$Pr[\mathcal{A}(m, L, SKT, \sigma)] > \frac{1}{n-t} + \frac{1}{Q(k)}. \tag{4.21}$$

The rest follows the same as the proof in ref. [130], we omit details here.

4.6.2.4 Efficiency Analysis of LRRS

To evaluate the complexity of our LRRS, we compare it with other related schemes in Table 4.2. Here, to simulate redaction, we derive a signing scheme directly from chameleon hash schemes [67,148,149]. As shown in Table 4.2, our LRRS is dominated by the size of the ring chosen for signing. Although our LRRS requires linear costs (with n) for each algorithm, these operations are acceptably efficient to run.

TABLE 4.2: Complexity of Signature Schemes

Schemes	Algorithms			
	Sign	Verify	Redact	Update
Our LRRS	$(5n-1)T_e$ $+(4n-1)T_m$	$(2n+2)T_e$ $+(3n+1)T_m$	$(4n+2)T_e$ $+(7n+1)T_m$	$2nT_e$ $+4nT_m$
Chen's CS-1 [148]	$3T_e + 1T_m + 1T_s$	$2T_e + 1T_m + 1T_v$	$2T_e + 1T_m + 1T_i$	n/a
Ateniese's CS [67]	$4T_e + 2T_m + 1T_s$	$2T_e + 1T_m + 1T_v$	$2T_e + 3T_m + 1T_i$	n/a
Chen's CS-2 [149]	$3T_e + 2T_m + 1T_s$	$2T_e + 2T_m + 1T_v$	$2T_e + 3T_m + 1T_i$	n/a

Denote CS as chameleon signature (signing directly on chameleon hash); n as number of ring members; T_m as group multiplication; T_e as group exponentiation; T_i as group inversion; T_p as bilinear pairing operation; n/a as not applicable; T_s as signing operation of BLS signature [31] signature; T_s as verification operation of BLS signature [31].

For further evaluation, we borrow results from our work [132]. We apply different key sizes and compare the tested algorithm with our work [150]. As shown in Figure 4.7, our LRRS scheme yields a much more stable cost in the key generation than our other in work [150]. As further demonstrated by Figures 4.8–4.11, each function of our LRRS is dominated by the factor of threshold n. As it is shown in Figure 4.12, within an approximately 100ms computing time, where we fix threshold n by 50, it generates 50 times of signing operations, 60 times of redaction and verification, 80 times of an update.

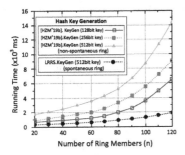

FIGURE 4.7: KeyGen Cost

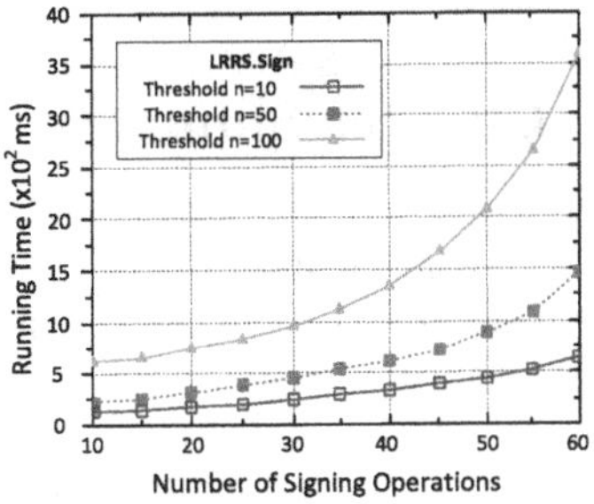

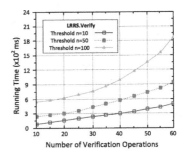

FIGURE 4.8: Signing Cost

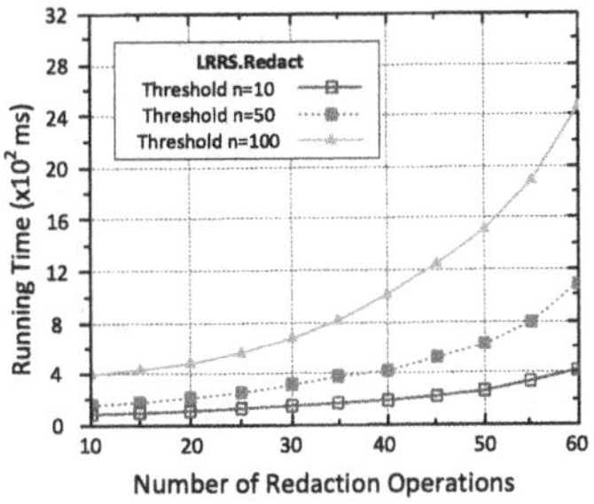

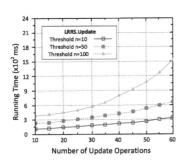

FIGURE 4.9: Redaction Cost

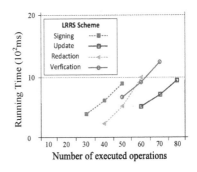

FIGURE 4.10: Verification Cost

FIGURE 4.11: Update Cost

FIGURE 4.12: Overview of Cost

4.7   Chapter Summary

The public key signature is a popular primitive used for the authentication. To authenticate a transaction , each user is assigned a public key for the verification and the private key signing. This chapter begins with an overview which outlines use cases and gives general background and definition for design. Next, it sequentially reviews and analyses the construction and security of ECDSA, BLS, and MlSAG schemes as case analysis. In the end, this chapter provides our work as an exercise based on previous studies. After reading this chapter, the reader should be able to: (1). Know how to define a public key signature scheme and its security model. (2). Understand the instantiated proving techniques by practicing the given examples. (3). Sketch the basic design of ECDSA, BLS, and MlSAG schemes, and evaluate the security via one of the proving techniques mentioned in this chapter. As concluded by this chapter, the design and analysis of a signature scheme follow by famous proving methodology and intrinsic empirical knowledge. By grasping the main ideas of these proof techniques and repetitive practices, the reader shall be able to eligible to devise a practical public key signature for blockchain and validate its security via the security reduction.

Chapter 5

Public-Key Encryption Scheme for Blockchain

5.1 Chapter Introduction

Public key encryption (PKE) is a vital tool to protect data confidentiality in the public channel. Researches of PKE have extended to various aspects: Identity-based, attribute-based, key management, key escrow, forward security, etc. While devising PKE for blockchain, academia accepted that IND-CCA2 is a practical security level for a proposed PKE to achieve. This chapter leverages the practical security and proof techniques to achieve practical designs and analysis of PKE schemes. This chapter begins with an overview of existing PKE schemes. Then, it systematically analyzes the design and security of two representative IND-CCA2 secure PKE schemes: Identity-Based Encryption (IBE) and Cramer and Shoup's encryption (CS) schemes. To exemplify the proof technique, it also gives an encryption scheme, which is IND-CPA secure. Next, it summarizes generic methods to achieve IND-CCA2 security.

After reading this chapter, you will:

- Know how to define public key encryption scheme and its security model.

- Understand the instantiated proving techniques by practicing on the given examples.

- Sketch the basic design of IBE and CS schemes and evaluate the security via one of the proving techniques mentioned in this chapter.

- Enumerate and explain generic methodologies to achieve IND-CCA2 security.

5.2   Overview of PKE

In this section, we give the first introduction to PKE. Then, we review PKE used in blockchain. After that, we give a formal definition for PKE.

5.3   Introduction to PKE

Diffie and Hellman [3] proposed the notion of public key infrastructure (PKI) in 1976, which started the era of modern cryptography. Their idea is to detach public key and private key so that users can use the public key for encryption and keep the private for decryption. The use of PKI enables the user to communicate securely through the public channel without establishing mutual trust in advance. PKI solves two major problems associated with symmetric encryption at a high level: (1) Key distribution, and (2) Digital signature. The former problem exists in symmetric encryption where two unfamiliar users have to negotiate a secret key for communication. The latter problem specifies the difficulty to determine the authenticity of received information in symmetric encryption setting.

Current PKE schemes can be categorized by following types of variants: Rabin algorithm [96] based on factorization, ElGamal algorithm [94] based on the discrete-logarithm problem in a finite field, and elliptic curve-based algorithms. To brief the history of PKE, early PKE only considers one-wayness for security. Briefly, it only seeks to prevent an adversary from recovering entire plaintext from a given ciphertext. However, as later identified in a stronger model, leakage of merely one-bit sensitive information hidden under the encryption is also insecure and leads to a security breach. Goldwasser and Micali [151,152] formalized this one-bit security by a security notion called Indistinguishability under Chosen-Plaintext Attack (IND-CPA). In 1990, Naor and Yung [153] introduced a stronger notion called Indistinguishability under Chosen-Ciphertext attack (IND-CCA1). In 1991, Rackoff and Simon proposed the strongest security notion Indistinguishability under Adaptive Chosen-Ciphertext attack (IND-CCA2). In practice, an encryption is generally required to achieve IND-CCA2 security.

5.3.1   PKE for Blockchain

Due to performance deficiencies of IND-CCA2 schemes, there are no representative encryption scheme with ideal security to be instantiated and implemented for blockchains. However, we review some standardized PKE schemes which are generally used, such as ECC-based encryption [154], RSA [93], and Pedersen commitment [155].

5.3.1.1 Use Case 1: ECC-Based Encryption in Bitcoin

Elliptic Curve Cryptography (ECC) is based on PKI. An elliptic curve is a plane algebraic curve defined by an equation of the form. ECC-based encryptions build on elliptic curve yield smaller key sizes with the same security as RSA. Due to their less intensive computations, it is ideal for mobile devices and networks. An ECC-based signature (ECDSA) has been first adopted in Bitcoin as the earliest use case of PKS in the blockchain. Since ECC schemes naturally yield encryption, one can easily adopt ECC-based encryption and implement it in Bitcoin with the same set of parameters initiated for the system.

5.3.1.2 Use Case 2: RSA in Blockchain

RSA [93] is a representative encryption scheme used in Bitcoin and blockchain. RSA [93] has been the most widely used PKE scheme in the world. The security of RSA relies on the practical difficulty of large integer factorization. To implement RSA, most blockchains follow a similar method for the creation and encryption of blockchain wallets. When creating a cryptocurrency wallet, a public address is assigned to the user who receives cryptocurrencies and encryption. The private key, on the other hand, will be used for spending cryptocurrencies as well as the decryption e.g. [1]. Though we do not instantiate RSA in this book, the reader can find plenty of resources to study RSA.

5.3.1.3 Use Case 3: Pedersen Commitments

Pedersen commitment [155] is a well-known variant scheme of the homomorphic commitment. Homomorphic commitment (or encryption) is widely used as building blocks in many cryptocurrencies, like RingCT [102], Zero-Coin [10], ZeroCash [46], Monero [44], etc. It allows committing a value without revealing it to the other parties by utilizing the homomorphic encryption technique. Basically, homomorphic commitment is a one-way trapdoor function (as discussed in Section 8.3.2) and considered as a simple but insecure version of encryption. The main reason is that, as discussed in Section 5.3 and Section 5.7, it generally implies an IND-CPA security which is not a satisfiable security (i.e., IND-CCA2 security) for practical application.

5.3.1.4 Other Cases

Generally, we consider any encryption schemes satisfying IND-CCA2 security as eligible for use in blockchains. This chapter will focus on designing and analyzing PKE schemes with IND-CCA2 security and generic methods to achieve IND-CCA2 security.

5.3.2   Definition of PKE

PKE is mainly used to achieve data confidentiality by encrypting every bit of the message. We then sketch the scenario of encryption and decryption of PKE as follows.

- Suppose Bob wishes to send a sensitive message m to Alice. Alice and Bob both join PKI to get a public/secret key pair (pk, sk).

- Bob inputs Alice's public key pk_A and message m to a PKE scheme. Bob derives a ciphertext c and sends it to Alice.

- On receiving c, Alice uses her private key sk_A to decrypt c and return m.

A public-key encryption scheme consists of the following four algorithms.

SysGen$_{\mathsf{PKE}}$: This algorithm takes as input a security parameter λ. It returns the system parameters SP.

KeyGen$_{\mathsf{PKE}}$: This algorithm takes as input SP. It returns a public/secret key pair (pk, sk).

Encrypt$_{\mathsf{PKE}}$: This algorithm takes as input a message m, pk, and SP. It returns a ciphertext $CT = E[SP, pk, m]$.

Decrypt$_{\mathsf{PKE}}$: This algorithm takes as input a CT, sk, and SP. It returns a message m or outputs $\perp$ to indicate a failure.

The validation of a PKE scheme generally consists of correctness and security analysis based on security requirement. We generalize them as follows:

Correctness. Given any (SP, pk, sk, m, CT), if $CT = E[SP, pk, m]$ is a ciphertext encrypted with pk on the message m, the decryption of CT with the secret key sk will return the message m.

Security. Without the secret key sk, it is hard for any PPT adversary to extract the message m (even a bit of information) from the given ciphertext $CT = E[SP, pk, m]$.

The security of a PKE scheme is modeled by a game played among a challenger and an adversary. Specifically, indistinguishability is a security notion of capturing the extraction of one-bit information from the ciphertext. The indistinguishability against chosen ciphertext attack (IND-CCA) requires that it is difficult for an adversary $\mathcal{A}$ to efficiently distinguish between a fresh ciphertext encrypted from two distinct messages while allowing adversary $\mathcal{A}$ to launch decryption queries adaptively.

Formally, we describe IND-CCA as follows.

Let SP be the system parameters.

Setup. The challenger runs SysGen$_{\mathsf{PKE}}$ and KeyGen$_{\mathsf{PKE}}$ to generate system parameters SP and a key pair (pk, sk). It then sends pk to the adversary $\mathcal{A}$. The challenger keeps sk to respond to decryption queries from $\mathcal{A}$.

Phase 1. $\mathcal{A}$ makes decryption queries on ciphertexts that are adaptively chosen by the adversary itself. For a decryption query on the ciphertext CT_i, the challenger runs $\mathsf{Decrypt}_{\mathsf{PKE}}$ and then sends the decrypted message to $\mathcal{A}$.

Challenge. $\mathcal{A}$ outputs two distinct messages m_0, m_1, which are adaptively chosen by $\mathcal{A}$ itself. The challenger randomly chooses $c \in \{0,1\}$ and then computes a challenge ciphertext $CT^* = E[SP, pk, m_c]$. It then returns it to $\mathcal{A}$.

Phase 2. The challenger responds to decryption queries in the same way as in Phase 1 with the restriction that no decryption query is allowed on CT^*.

Guess. $\mathcal{A}$ outputs a guess c' of c and wins the game if $c' = c$.

5.4 Case Analysis: IBE

In this section, we review Boneh et al.'s [156] Identity-Based Encryption (IBE) under random oracle model. Our review consists of IBE's construction and security analysis.

5.4.1 Background of IBE Scheme

In 1984, Shamir [157] proposed notion of identity-based encryption (IBE). This is a different infrastructure for encryption in comparison with traditional public key infrastructure (PKI) [158], which adopts a unique identifier as the public key. In 2001, Boneh and Frankly [156] proposed the first practical IBE scheme. After that, researches on IBE have accelerated. To achieve practical security without the assumption, Boneh et al. proposed two secure IBE schemes provably secure in standard model [159,160]. Then, in 2005, Waters [161] proposed a practical IBE scheme with ideal overall performance. Similarly, in 2006, Gentry [162] proposed a practical IBE scheme. Specifically, Gentry's proposal is the tight security reduction in the standard model.

Achieving tight security reduction based on strong security definition and weak intractability assumption in the standard model is challenging and meaningful work. Also, it is worth investigating the ciphertext with the short length and low costs. More details can be found in ref. [163].

Currently, there are some attempts to introduce identity-based encryption or infrastructure to blockchain [164–166]. For example, Xue et al. [165] propose an identity-based public auditing (IBPA) scheme for cloud storage systems. Zhou et al. [164] also propose an improved key distribution solution by exploiting blockchain and identity-based encryption. These works mainly start from exploiting benefits of identity-based infrastructure, yet not from the practical security perspective for blockchain.

5.4.2   Construction of IBE Scheme

The notion of identity-based encryption (IBE) is different from public key infrastructure, as it adopts a unique identifier (e.g. user's driver licence) to assure the user's identity for the public key encryption. By knowing the recipient's public key, the sender is not required to query an online certificate authority (CA) [167] about the recipient's public key.

Boneh et al. [156] proposed the first practical IBE scheme. For ease of presentation and analysis, Boneh et al. build their scheme by three steps: BasicPub, BasicIdent, and FullIdent. The last version: FullIdent is their complete construction of the IBE scheme. Following this idea, we briefly review the sketch of each scheme.

5.4.2.1   BasicPub Scheme

BasicPub Scheme of IBE	
Public key: $< q, G_1, G_2, \hat{e}, n, P, P_{pub}, Q_{ID}, H_2 >$; Private key: $d_{ID} = sQ_{ID} \in \mathcal{G}_1^*$	
Encrypt	**Decrypt**
1. Given $M \in \{0,1\}^n$, pick a random $r \in \mathbb{Z}_q^*$.	4. Given $C = < U, V >$, compute plaintext $M = V \oplus H_2(\hat{e}(d_{ID}, U))$
2. Compute $g = \hat{e}(Q_{ID}, P_{pub}) \in G_2^*$.	
3. Compute ciphertext $C = < rP, M \oplus H_2(g^r) >$	

FIGURE 5.1: Sketch of BasicPub Scheme

To start, Boneh and Franklin firstly propose BasicPub toward their IBE proposal. We give a sketch of this basic PKE scheme in Figure 5.1.

5.4.2.2   BasicIdent Scheme

Boneh and Franklin improve BasicPub via the BasicIdent. Here, BasicIdent is an utter IBE scheme characterized by an **Extract** algorithm to derive a private key for each user. We give a sketch of this basic IBE scheme in Figure 5.2.

BasicIdent Scheme of IBE

System parameter: $param_{BasicIdent} = <q, G_1, G_2, \hat{e}, n, P, P_{pub}, H_1, H_2>$; Master key is $s \in Z_q^*$

Encrypt	Decrypt
1. Given $ID \in \{0,1\}^*$, compute public key: $Q_{ID} = H_1(ID) \in G_1^*$, private key $d_{ID} = sQ_{ID}$.	5. Given $C = <U, V>$, compute plaintext
2. Given $M \in \mathcal{M}$, compute $Q_{ID} = H_1(ID) \in G_1^*$.	$M = V \oplus H_2(\hat{e}(d_{ID}, U))$
3. Pick a random $r \in Z_q^*$.	
4. Compute $g_{ID} = \hat{e}(Q_{ID}, P_{pub} \in G_2^*)$. Derive ciphertext $C = < rP, M \oplus H_2(g_{ID}^r) >$	

FIGURE 5.2: Sketch of BasicIdent Scheme

5.4.2.3 FullIdent Scheme

Boneh and Franklin ultimately construct an IND-CCA2 secure IBE scheme FullIdent from the basic version BasicIdent. We give a sketch of this final version in Figure 5.3.

FullIdent Scheme of IBE

System parameter: $param_{BasicIdent} = <q, G_1, G_2, \hat{e}, n, P, P_{pub}, H_1, H_2, H_3, H_4>$; Master key is $s \in Z_q^*$

Encrypt	Decrypt
1. Given $ID \in \{0,1\}^*$, compute public key: $Q_{ID} = H_1(ID) \in G_1^*$, private key $d_{ID} = sQ_{ID}$.	7. Given $C = < U, V, W >$, check whether $U \notin G_1^*$
2. Given $M \in \mathcal{M}$, compute $Q_{ID} = H_1(ID) \in G_1^*$.	8. If yes, proceed; otherwise, reject and output $\perp$.
	9. Compute $V \oplus H_2(\hat{e}(d_{ID}, U)) = \sigma$.
3. Pick a random $\sigma \in \{0,1\}^n$.	10. Compute $W \oplus H_4(\sigma) = M$.
4. Compute $r = H_3(\sigma, M)$.	11. Set $r = H_3(\sigma, M)$.
5. Compute $g_{ID} = \hat{e}(Q_{ID}, P_{pub}) \in G_2$.	12. Check whether $U = rP$. If not, reject and
6. Derive ciphertext $C = < rP, \sigma \oplus H_2(g_{ID}^r), M \oplus H_4(\sigma) >$	output $\perp$; otherwise, output M.

FIGURE 5.3: Sketch of FullIdent Scheme

5.4.3 Security Analysis of IBE Scheme

The proof of Boneh and Franklin's FullIdent scheme consists of the security of the scheme and reduction. Following this idea, we give an analysis of each scheme next.

5.4.3.1 Security Analysis BasicPub

Boneh and Franklin proved that BasicPub scheme is IND-CPA secure if the BDH assumption holds in Lemma 4.3 in ref. [156]. Since it is simple and can be easily interpreted, we omit details here but focusing on the security of BasicIdent and FullIdent.

5.4.3.2   Security Analysis of BasicIdent

Proof Overview 5.1 The security analysis of BasicIdent is a building block for security of FullIdent. Here, we only give a proof sketch of BasicIdent. The security proof of BasicIdent follows Theorem 5.1 and Lemmas 5.1, 5.2. We simplify Theorem 5.1, Lemmas 5.1 and 5.2 for ease of understanding. The proof of Lemmas 5.1 and 5.2 implies Theorem 5.1. In other words, if we can prove Lemmas 5.1 and 5.2, Theorem 5.1 is proved. Since Lemma 5.1 is proved by Fujisaki and Okamoto in ref. [168], it remains to prove Lemma 5.2. However, since the proof of Lemma 5.2 is similar to the proof of Lemma 5.4, we analyze it in Section 5.4.4, and omit details here. We recommend the reader leave this session as a future practise, and skip directly to Section 5.4.4. After interpreting how to prove Lemma 5.4, the reader shall easily finish the proof of Theorem 5.1.

To help reader form a concept of proof for Lemmas 5.1 and 5.2, we sketch the workflows of game modeled for Lemmas 5.1 and 5.2 in Figures 5.4 and 5.5, respectively.

Theorem 5.1 *BasicIdent is IND-ID-CPA secure if BDH assumption holds.*

Lemma 5.1 *IND-ID-CPA attack on BasicIdent can be converted to an IND-CPA attack on BasicPub.*

Lemma 5.2 *BasicPub is IND-CPA secure if the BDH assumption holds.*

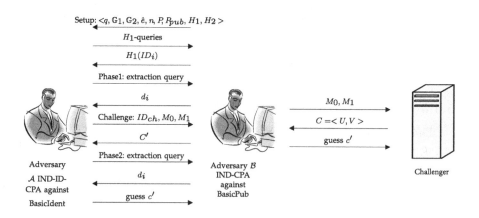

FIGURE 5.4: The Sketch of Theorem 5.1

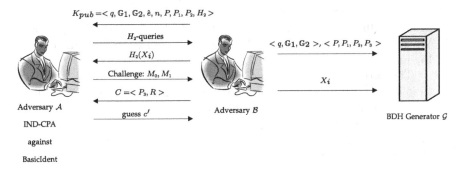

FIGURE 5.5: The sketch of Lemma 5.1

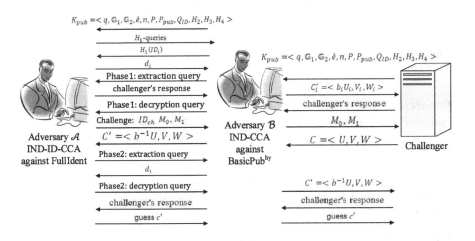

FIGURE 5.6: The Sketch of Lemma 5.2

5.4.4 Security Analysis of FullIdent

Proof Overview 5.2 To prove security of IBE, Boneh and Franklin give Theorem 5.2 to define all the preconditions of a FullIdent scheme to achieve IND-ID-CCA security. Theorem 5.2 defines corresponding advantages of an adversary $\mathcal{A}$ against FullIdent. Our goal in this part is to prove Theorem 5.2.

In order to prove Theorem 5.2, security analysis breaks into three parts: Lemma 5.3, Theorem 5.3, Lemma 5.4, and corresponding proof. This division is such that each part yields a trivial analysis of an independent scheme that is easy to discuss and prove. As a result, by asserting these three parts, we can trivially reduce an IND-ID-CCA adversary against FullIdent to an efficient algorithm which solves the BDH problem. We give a sketch of Theorem 5.1 in Figure 5.4, a sketch of Lemma 5.1 in Figure 5.5, sketch of Lemma 5.2 in Figure 5.6 respectively.

The proof details of Lemma 5.3 are given in [156]. Since it is easy to follow, we omit details here. Also, Fujisaki and Okamoto prove Theorem 5.3 in their work [168]. Specifically, they show how to transform an IND-CCA adversary against $BasicPub^{hy}$ (a hybrid encryption scheme following their design [168]) to an IND-CPA adversary against BasicPub. We also omit the details.

We mainly focus on analyzing the proof of Lemma in this part. The main proof technique used here is proof by transitions of failure events (discussed in Section 2.4.2.2), case analysis, and reduction based on previous work. A sketch of Lemma 5.4 is shown in Figure 5.6.

Similarly, we simplify definitions of Theorem 5.2, Lemma 5.3, Theorem 5.3, and Lemma 5.4 for ease of understanding. Refer to ref. [156] for more details.

Theorem 5.2 *Suppose there is an efficient IND-ID-CCA adversary $\mathcal{A}$ against FullIdent scheme. Then, there is a BDH algorithm $\mathcal{B}$ for $\mathcal{G}$:*

To prove Theorem 5.2, it immediately follows Lemma 5.3, Theorem 5.3, and Lemma 5.4 defined as below.

Lemma 5.3 *If $\mathcal{A}$ is an efficient IND-CPA adversary against BasicPub, an algorithm $\mathcal{B}$ solves the BDH problem efficiently.*

Theorem 5.3 *If $\mathcal{A}$ is an efficient IND-CCA adversary against BasicPub, there is an efficient IND-CPA adversary against BasicPub.*

Lemma 5.4 *If $\mathcal{A}$ is an efficient IND-ID-CCA adversary against FullIdent, there is an efficient adversary against BasicPub.*

Tip 5.1 As observed, Lemma 5.3 shows IND-ID-CCA attack can be converted into IND-CCA attack. Theorem 5.3 shows IND-CCA adversary implies an IND-CPA adversary on BasicPub. Lemma 5.3 shows IND-CPA adversary implies an algorithm for BDH. Therefore, an IND-ID-CCA adversary against FullIdent implies an algorithm $\mathcal{B}$ that solves the BDH problem with reduction in $(\mathcal{O}(time\mathcal{A}), 2\epsilon(k)/q_{H_2})$, where Lemma 5.4, Theorem 5.3, and Lemma 5.3 can be proved.

Further, Theorem 5.3 is proved by Fujisaki and Okamoto in ref. [168]. As the proof of Lemma 5.3 can be easily interpreted by ref. [156], we omit details here.

For ease of understanding, we show the sketch of Lemma 5.4 in Figure 5.6 (i.e. constructing an IND-CCA adversary $\mathcal{B}$ that uses $\mathcal{A}$ to gain advantage $\frac{\epsilon}{e(1+q_E+q_D)}$ against $BasicPub^{hy}$).

Tip 5.2 Following Figure 5.6, we can easily observe the workflow of proof of Lemma 5.4. As depicted by Figure 5.6, a challenger is employed to simulate $BasicPub^{hy}$ scheme. It interacts with adversary $\mathcal{B}$ in order to derive an answer to BDH instance. Meanwhile, adversary $\mathcal{A}$ tries to simulate a challenger for an IND-ID-CCA adversary $\mathcal{A}$ against FullIdent scheme. $\mathcal{B}$ seeks to exploit $\mathcal{A}$'s guess c' and sends to the challenger to win the game.

As shown in Figure 5.6, the right side is a typical game for IND-CCA game, while the left side is a IND-ID-CCA game featured by extraction query for the user's identity. The tricky part is the design of C'. The reader may ask why do we define $C' =< b^{-1}U, V, W >$? We simply answer this question by "making successful decryption of challenged ciphertext for two different schemes". Concretely, this design allows for decryption of C' using d_{ch} in BasicIdent that equals to decryption of C using d_{ID} using Basic Pub. Let's see details as below.

Proof of Lemma 5.4 Following Figure 5.6, we prove Lemma 5.4 by showing how to construct a game between the challenger and the adversary.

The game between the challenger and the adversary $\mathcal{B}$ starts with the challenger first generating a random public key by running the algorithm KeyGen of BasicPub^{hy}.

The result is a public key $K_{pub} =< q, \mathcal{G}_1, \mathcal{G}_2, \hat{e}, n, P, P_{pub}, Q_{ID}, H_2, H_3, H_4 >$ and a private key $d_{ID} = sQ_{ID}$. The challenger gives K_{pub} to $\mathcal{B}$.

$\mathcal{B}$ interacts with $\mathcal{A}$ to launch an IND-CCA attack on the key K_{pub} as follows:

Setup: $\mathcal{B}$ gives $\mathcal{A}$ the system parameters of BasicIdent $< q, \mathcal{G}_1, \mathcal{G}_2, \hat{e}, n, P, P_{pub}, Q_{ID}, H_2, H_3, H_4 >$.

H_1-**queries:** $\mathcal{A}$ can submit queries the random oracle H_1. To respond to these queries $\mathcal{B}$ keeps a list of tuples $< ID_j, Q_j, b_j, c_j >$, denoted as H_1^{list}. When $\mathcal{A}$ queries the oracle H_1 at a point ID_i, $\mathcal{B}$ proceeds as follows:

1. If the query ID_i already exists on the H_1^{list} in a tuple $< ID_i, Q_i, b_i, c_i >$, $\mathcal{B}$ responds with $H_1(ID_i) = Q_i \in \mathbb{G}_1^*$.

2. Otherwise, $\mathcal{B}$ selects a random $coin \in \{0,1\}$, so that $\Pr[coin = 0] = \delta$ for some δ that will be determined later.

3. $\mathcal{B}$ picks a random $b \in Z_q^*$.

   If $coin = 0$, compute $Q_i = bP \in G_1^*$. If $coin = 1$, compute $Q_i = bQ_{\text{ID}} \in G_1^*$.

4. $\mathcal{B}$ adds the tuple $< ID, Q, b, c >$ to the H_1^{list} and responds to $\mathcal{A}$ with $H_1(ID_i) = Q_i$.

Phase 1: Private key queries. $\mathcal{B}$ responds to this query as follows:

1. Run H_1-queries to obtain a $Q_i \in G_1^*$ such that $H_1(ID_i) = Q_i$. Let $< ID_i, Q_i, b_i, c_i >$ be the corresponding tuple on the H_1^{list}. If $coin_i = 1$, then $\mathcal{B}$ reports failure and terminates. Attack fails.

2. If $coin_i = 0$, hence $Q_i = b_i P$. Define $d_i = b_i P_{pub} \in G_1^*$. Return d_i to $\mathcal{A}$.

Phase 1: Decryption queries. $\mathcal{B}$ responds as follows:

1. Run H_1-queries to obtain a $Q_i \in G_1^*$ such that $H_1(ID_i) = Q_i$.

2. Suppose $coin_i = 0$. Run private key queries to respond to the decryption query.

3. Suppose $coin_i = 1$. Then $Q_i = b_i Q_{\text{ID}}$. Set $C_i' =< b_i U_i, V_i, W_i >$. Relay $< C_i' >$ to the challenger and relay the challenger's response back to $\mathcal{A}$.

Challenge: $\mathcal{A}$ outputs ID_{ch} and M_0, M_1. $\mathcal{B}$ proceeds as follows:

1. $\mathcal{B}$ gives the challenger M_0, M_1. The challenger responds with $C =< U, V, W >$.

2. $\mathcal{B}$ runs H_1-queries to obtain a $Q \in G_1^*$ such that $H_1(ID_{ch}) = Q$. Let $< ID_{ch}, Q, b, coin >$ be the corresponding tuple on the H_1^{list}. If $coin = 0$, then $\mathcal{B}$ reports failure and terminates. The attack failed.

3. If $coin = 1$ and therefore $Q = bQ_{\text{ID}}$. Set and return $C' =< b^{-1}U, V, W >$ to $\mathcal{A}$.

Phase 2: Private key queries. Same as **Phase 1**.

Phase 2: Decryption queries. Same as **Phase 1**. If relaying a query equal to the challenge ciphertext $C =< U, V, W >$, then reports failure.

Guess: $\mathcal{A}$ outputs a guess c' for c. Algorithm $\mathcal{B}$ outputs c' as its guess for c.

Tip 5.3 It remains to bound the adversary $\mathcal{A}$'s winning probability in the above game. Similar as we previously introduced in Section 4.4.3 for analyzing the BLS scheme's security, we deal with the abortion case of game and discuss the winning case by presenting it in the algebraic format. Since this part is similar to security analysis of BLS scheme, we omit details here.

Claim 5.1 *If algorithm $\mathcal{B}$ does not abort during the simulation, then algorithm $\mathcal{A}$'s view is identical to its view in the real attack. Furthermore, if $\mathcal{B}$ does not abort, then $|\Pr[c = c'] - \frac{1}{2}| \geq \epsilon$.*

Proof of claim 5.1 As observed, the responses to H_1-queries are as in the real attack since each response is uniformly and independently distributed in G_1^*. All responses to private key extraction queries and decryption queries are valid. Finally, the challenge ciphertext C' given to $\mathcal{A}$ is the FullIdent encryption of M_c for some random $c \in \{0,1\}$. Therefore, by definition of algorithm $\mathcal{A}$, we have $|\Pr[c = c'] - \frac{1}{2}| \geq \epsilon$.

To bound the probability of algorithm $\mathcal{B}$ during the simulation, we do a case analysis on the reasons for abortion:

1. bad private key query,

2. bad ID_{ch} to be challenged on,

3. bad decryption query.

Further, we can define three events and analyze them accordingly to bound the $\mathcal{B}$'s probability without abortion which is at least $\frac{1}{e(q_E + q_D + 1)}$. Refer to the proof of Lemma 4.6 in ref. [156].

5.5 Case Analysis of CS

In this section, we review IND-CCA2 secure encryption scheme proposed by Cramer and Shoup (CS) [169]; we denote it as CS scheme. Specifically, we review the basic construction and security analysis of the CS scheme. To note, CS is the first practically secure IND-CCA2 scheme as it does not rely upon an assumption of the random oracle but based on the standard model.

5.5.1 Construction of CS

We sketch the basic construction of the Cramer and Shoup's encryption [169] in Figure 5.7. As the construction of CS scheme is simple from an observation, we omit detailed discussion here.

Cramer and Shoup's Encryption
Public key: $\text{param}_{\text{CS}} = \{G, q, g_1, g_2, c, d, h, H\}$; Private key is $\{x_1, x_2, y_1, y_2, z\}$

Encrypt	Decrypt
1. Given a message $m \in G$, pick a random $r \in Z_q$.	6. Given a ciphertext (u_1, u_2, e, v), compute $\alpha = H(u_1, u_2, e)$.
2. Compute $u_1 = g_1^r$ and $u_2 = g_2^r, e = h^r m$.	7. Check whether $u_1^{x_1 + y_1 \alpha} u_2^{x_2 + y_2 \alpha} = v$.
3. Compute $\alpha = H(u_1, u_2, e)$.	8. If yes, proceed; otherwise, reject and output $\perp$.
4. Compute $v = c^r d^{r\alpha}$.	9. Compute plaintext $m = e/u_1^z$.
5. Derive ciphertext (u_1, u_2, e, v)	

FIGURE 5.7: Sketch of Cramer and Shoup's Encryption Scheme

5.5.2   Security Analysis of CS

Proof Overview 5.3 Proving techniques used in this work are proof by contradiction, case analysis, and proof by the theorem. There exists some mathematical terms (e.g. hyperplane) and concept (e.g. line $\mathcal{L}$ formed by intersecting the hyperplanes) which are difficult to understand. This also makes programming security reduction under the standard model-based cryptographic scheme non-trivial. Therefore, learners need to take every opportunity to learn the proving technique for this scheme.

At a high level, the proof of CS scheme implies a statistical test which distinguishes R from D. This test can be utilized as an efficient solution for DDH problem. To note, distribution R defines a set of random tuples where $(g_1, g_2, u_1, u_2) \in G^4$ while distribution D defines tuple such that $(g_1, g_2, u_1, u_2) \in G^4$ where $u_1 = g_1^r$ and $u_2 = g_2^r$.

Theorem 5.4 *The CS scheme [169] is secure against adaptive chosen ciphertext attack assuming that (1) the hash function H is chosen from a universal one-way family, and (2) the Diffie-Hellman decision problem is hard in the group G.*

Tip 5.4 As observed from Theorem 5.4, CS does not rely on the assumption of the random oracle. This is essential to build practical security, but it also makes the proof harder. To prove Theorem 5.4, we immediately assume an adversary against the proposed cryptosystem (namely, CS scheme). This is a typical proving technique called Proof by contradiction.

The proof begins by defining the simulator, decryption query, and encryption oracle.

To prove the Theorem 5.4, we will assume that there is an adversary $\mathcal{A}$ that can break the CS scheme [169] that the hash family is universal one-way. Specifically, we will focus on showing how to use this adversary to construct a reduction to reduce adversary's attack on solving the underlying intractable assumption. Here, the reduction results in a statistical test for the decisional Diffie-Hellman problem, DDHP. Following the definitions for DDKP given in Section 2.2 in ref. [169], we next show how to programme reduction from adversary's attack to the solution to DDHP.

Tip 5.5 Following the security model formalized in Section 5.3.2 for IND-CCA, tuple (g_1, g_2, u_1, u_2) is given from either the distribution **R** or **D**. We build a simulator that simulates the joint distribution consisting of adversary's $\mathcal{A}$ view in its attack, and the hidden bit b. Ideally, we will show that the adversary neither has a negligible advantage in guessing the hidden bit b nor has dependent view on b from **D** or **R**, respectively. To generalize, the above implies a statistical test distinguishing **R** from **D**. Briefly, run the simulator and adversary together, and if the simulator outputs b and the adversary outputs b', the distinguisher outputs 1 if $b = b'$, and 0 otherwise. The construction of such a simulator is as follows:

---

1. On given a (g_1, g_2, u_1, u_2), the simulator runs key generation algorithm with g_1, g_2.

2. The, the simulator randomly picks $x_1, x_2, y_1, y_2, z_1, z_2 \in Z_q$ and computes $c = g_1^{x_1} g_2^{x_2}, d = g_1^{y_1} g_2^{y_2}$, and $h = g_1^{z_1} g_2^{z_2}$.

3. The adversary is given (g_1, g_2, c, d, h, H). The simulator controls $(x_1, x_2, y_1, y_2, z_1, z_2)$. Adversary $\mathcal{A}$ set $z_2 = 0$.

4. The simulator answers decryption queries with $m = e/(u_1^{x_1} u_2^{z_2})$.

5. Given m_0, m_1, the simulator randomly chooses $b \in \{0, 1\}$ and computes

$$ e = u_1^{z_1} u_2^{z_2} m_b, \quad \alpha = H(u_1, u_2, e), \quad v = u_1^{x_1+y_1\alpha} u_2^{x_2+y_2\alpha}. \tag{5.1} $$

6. The simulator outputs (u_1, u_2, e, v).

Lemma 5.5 and 5.6 are essential in order to derive Theorem 5.4. Here, we give a simplified version.

Lemma 5.5 *When the simulator's input comes from* **D**, *the joint distribution of the adversary's view and the hidden bit b are statistically indistinguishable from that in the actual attack.*

Proof of Lemma 5.5. By observing the computation of u_1, the reader shall be clear that the output of the encryption and the decryption oracles have the ring distribution. Consequently, the lemma follows immediately from the following claim:

Claim. The decryption oracle in both an actual attack and an attack against the simulator rejects all invalid ciphertexts, except with negligible probability.

To prove this claim, one shall consider the distribution of the point $P = (x_1, x_2, y_1, y_2) \in Z_q^4$. To start, consider P as a random point on the plane $\mathcal{P}$ formed by intersecting the hyperplanes from the adversary's view. Then, we have:

$$\log c = x_1 + wx_2, \tag{5.2}$$

$$\log d = y_1 + wy_2. \tag{5.3}$$

However, due to the design of encryption oracle, we have:

$$\log v = rx_1 + wrx_2 + \alpha ry_1 + \alpha rwy_2. \tag{5.4}$$

Finally, consider invalid ciphertext which will not be rejected. We have:

$$\log v' = r_1'x_1 + wr_2'x_2 + \alpha'r_1'y_1 + \alpha'r_2'wy_2. \tag{5.5}$$

Based on **Eq 5.1**, **Eq 5.2**, and **Eq 5.4**, and their linearly independence, one can conclude decryption oracle's rejection by probability $1 - 1/q$. This suffices to complete the proof.

Tip 5.6 The above completes the proof of Lemma 5.5. At a high level, the proof of Lemma 5.5 is proceeded by cases analysis and induction. Here, the induction is highly logical and uneasy to follow. Additionally, a claim is given to capture the rejection of decryption oracle. The proof of claim leads to discussing the distribution of point P formed by intersecting the hyperplanes. To explore, Cramer and Shoup induct considering equations 5.2, 5.3, and 5.5 which yields a line resulted from $\mathcal{H}$ intersecting $\mathcal{P}$. Here, it is hard for beginners to induct from equations 5.2, 5.3, and 5.5 to a line result from the intersection by hyperplane $\mathcal{H}$ and plane $\mathcal{P}$. This requires some basic concept or intuition of geometry which is not essentially equipped by the cryptographic learner (especially those without mathematics basics).

Lemma 5.6 *When the simulator's input comes from* **R***, the distribution of the hidden bit b is independent of the adversary's view.*

The proof of Lemma 5.6 follows from the following two claims.

Claim 1. If the decryption oracle rejects all invalid ciphertexts during the attack, then the distribution of the hidden bit b is independent of the adversary's view.

The proof of Claim 1 mainly proceeds by discussing the information leakage of random point $\mathbf{Q}$. From illation, one obtains the following two equations:

$$\log h = z_1 + wz_2, \tag{5.6}$$

$$\log \epsilon = r_1 z_1 + wr_2 z_2. \tag{5.7}$$

Based on the linearly independent **Eqs 5.5** and **5.6**, one can conclude the conditional distribution of ϵ as a perfect one-time pad. This suffices to complete the proof.

Tip 5.7 Lemma 5.6 tends to discuss the distribution of b for input from R. In order to discuss the behavior of the decryption oracle, Claim 2 is given. The proof of Claim 2 is as follows: Cramer and Shoup first assume a point $Q = (z_1, z_2)$. By inferring the point Q and the output from the encryption oracle, they can derive two equations. By further illation, proof of Claim 2 completes. As we can see, illation is a basic skill for analysis which can be acquired by practices.

Claim 2. The decryption oracle will reject all invalid ciphertexts, except with negligible probability.

The proof of Claim 2 is mainly proceeded by the case analysis. First of all, capturing a random point P in the adversary's view allows one to conclude:

$$\log v = r_1 x_1 + wr_2 x_2 + \alpha r_1 y_1 + \alpha wr_2 y_2 \tag{5.8}$$

Then, assume the submission of an invalid ciphertext $(u_1', u_2', e', v') \neq (u_1, u_2, e, v)$, and proceed the discussion as case analysis according to the equalities of (u_1', u_2', e'), (u_1, u_2, e), α, and α', one shall complete the proof. We leave the case analysis as the future exercise for the reader. Refer to ref. [169] for more details.

Tip 5.8 To prove Lemma 5.6, Claim 2 is given to define the decryption oracle. To prove Claim 2, we follow the proof of Claim 1's pattern. First, assume a point $P = (x_1, x_2, y_1, y_2)$. Then, consider the response of the encryption and the decryption oracles. Further, consider three cases for analyzing an equation with as input. The proof technique used here is case

---

analysis. The tricky part here is to compute determinant in Case 2. This requires basic algebraic skill to handle the computation of determinant. The rest of case analysis is proceeded by careful illation.

5.6   Case Analysis of HDRS

This section presents our original work, a homomorphic and designated receiver signcryption (HDRS), as an example work. However, our HDRS achieves weaker security than INC-CCA, namely, IND-CPA. However, this work can help the reader practice and take their first steps to propose a new scheme.

5.6.1   Background of Signcryption

Signcryption is a primitive to achieve encryption and signature simultaneously in a single operation. As a result, data confidentiality and authentication are achieved simultaneously. Zheng [170] proposed the first signcryption scheme based on ElGamal signature and encryption in 1997. Then, Baek et al. [171] formally proved the confidentiality of Zheng's signcryption scheme [170] under a rigorous security model. Current signcryption schemes can be classified by: Public key infrastructure-based (PKI-based) [172–174] and Identity-based (ID-based) [175–189].

To summarize, we compare relevant schemes in Table 5.1. As it is shown in Table 5.1, almost half of them are provably secure under the standard model.

5.6.2   Construction of HDRS

We give a sketch of construction for HDRS in Figure 5.8. We give a short description of each algorithm. It suffices for the reader to perform security analysis. The reader is also referred to as ref. [190] for more details.

TABLE 5.1: Overview of Signcryption Schemes

Scheme	Identity-based	Online/Offline	Assumption	Security in Standard Model
IOSL [181]	Yes	Yes	q-SDHP and q-BDHIP	No
OOIB [182]	Yes	Yes	l-BDHI and l-SDHP	No
IBOO [183]	Yes	Yes	k-mBDHIP	No
EIOE [184]	Yes	Yes	k-CCA1	No
AISI [185]	Yes	No	CDHP and DBDH	Yes
SISS [186]	Yes	No	CDHP and DBDH	Yes
FSIS [187]	Yes	No	CDHP and DBDH	Yes
IBSS [188]	Yes	No	CDHP and DBDH	Yes
PSIB [189]	Yes	No	MBSDH and DMBDHI	No

Huang et al.'s HDRS Encryption

System parameter: $param_{HDRS} = \{G_2, p, g, H_2\}$;
Private key: $sk_{(\cdot)} = \{x_{0,(\cdot)}, x_{1,(\cdot)}, x_{2,(\cdot)}\}$
Public key: $pk_{(\cdot)} = \{y_{0,(\cdot)} = g^{x_{0,(\cdot)}}, y_{1,(\cdot)} = g^{x_{1,(\cdot)}}, y_{2,(\cdot)} = g^{x_{2,(\cdot)}}\}$.
Parse $(\cdot) = S$ or R, i.e., sender or receiver.

Signcrypt	Designcrypt
1. Given $CID \in \{0,1\}^*$ and $m \in Z_p$, compute $e = H_2(CID, pk_R)$ and $h = g^e$.	5. Given a signcrypted message $c = (c_0, c_1, c_2, c_3)$, compute $e = H_2(CID, pk_R)$.
2. Pick a random $\alpha \xleftarrow{R} Z_p^*$.	6. Decrypt as $m = \log_g \frac{c_2}{c_0^{(e+x_{0,R})}}$.
3. Compute $c_0 = g^\alpha$, $c_1 = y_{1,S}^\alpha$, $c_2 = g^m(h \cdot y_{0,R})^\alpha$, $c_3 = (y_{1,R}^m \cdot y_{2,R}^\alpha)^{x_{1,S}}$.	7. Check whether $c_3 = y_{1,S}^{m \cdot x_{1,R}} \cdot c_1^{x_{2,R}}$.
4. Derive a signcrypted ciphertext $c = (c_0.c_1, c_2, c_3)$.	8. If yes, accept; otherwise, reject.

FIGURE 5.8: Sketch of HDRS Scheme

HDRS.Setup$(\lambda) \to (param_{LRRS})$. Denote λ as a security parameter.
HDRS.KeyGen$(param_{LRRS}) \to (sk_{user}, pk_{user})$
HDRS.RKeyGen$(param_{LRRS}, sk_{S_A}, sk_{S_B}) \to (k_{AB})$
HDRS.Signcrypt$(CID, m, sk_S, pk_R) \to (c, \perp)$.
Parse c as:

$$c_0 = g^\alpha, \tag{5.9}$$

$$c_1 = y_{1,S}^\alpha, \tag{5.10}$$

$$c_2 = g^m(h \cdot y_{0,R})^\alpha, \tag{5.11}$$

$$c_3 = (y_{1,R}^m \cdot y_{2,R}^\alpha)^{x_{1,S}}. \tag{5.12}$$

HDRS.Re-Sign$(c_{S_A}, k_{AB}) \to (c_{S_B})$

HDRS.De-Signcrypt$(c, \mathsf{CID}, sk_R) \to (m)$. Parse designcryption as:

$$m = \log_g^{c_0 \overline{\frac{c_2}{(e+x_{0,R})}}}. \tag{5.13}$$

HDRS.Verify$(m, \mathsf{CID}, c, sk_R) \to (0 \text{ or } 1)$.

5.6.3   Security Requirement of HDRS

Here, we define the security requirement of our HDRS to achieve confidentiality (i.e. indistinguishability of ciphertext). To explain, indistinguishability against chosen-plaintext-attacks (IND-CPA) is weaker security than IND-CCA in the sense that adversary cannot launch decryption queries. Specifically, it is captured by experiment $IND - CPA_{\mathcal{A}}^{\mathsf{HDRS}}(\lambda)$. Assume there exists an efficient adversary $\mathcal{A}$ against our scheme, our HDRS is secure if $\Pr| \left[IND - CPA_{\mathcal{A}}^{\mathsf{HDRS}}(\lambda) = 1 \right] - \frac{1}{2} | \leq \nu(\lambda)$ holds where ν is a negligible function.

Experiment: $IND - CPA_{\mathcal{A}}^{\lambda}$

    $param_{\mathsf{LRRS}}^{ch} \leftarrow \mathsf{HDRS.Setup}(\lambda)$
    $(pk_R, sk_R)(pk_S, sk_S) \leftarrow \mathsf{HDRS.KeyGen}(param_{\mathsf{TUCH}}^{ch})$
    $b^* \leftarrow \mathcal{A}^{\mathsf{Encrypt}}(pk_R, pk_S, c_b, m_0, m_1)$
    where oracle Encrypt on input: $pk_R, \mathsf{CID}, m, \alpha$:
    return $C = g^m (h \cdot pk_R)^\alpha$, i.e.
    the same way as HDRS.Signcrypt did for c_2
    in a signcrypted message $c = (c_0, c_1, c_2, c_3)$
    Here, $C_b = g^{m_b} (h \cdot pk_R)^\alpha$
    where b is chosen randomly from $\{0,1\}$
       out of $\mathcal{A}$'s view and α is chosen
    randomly from Z_q and $|m_0| = |m_1|$
    Return 1 if $b = b^*$
    else, return 0.

5.6.4   Security Analysis of HDRS

Proof Overview 5.4 We prove the security of HDRS scheme mainly by transitions based on indistinguishability. The proof is programmed so that Game 0 and Game 1, Game 1 and Game 2 are indistinguishable (it only makes negligible changes); otherwise, the distinguishability results in a solution to DDHP. The proof of this scheme is obvious.

Privacy: Assume $\mathcal{A}$ is a PPT adversary $\mathcal{A}$ who can break the IND-CPA security of our HDRS. We first prove by game hopping where each hop only changes $\mathcal{A}$'s view negligibly. We can then construct an algorithm $\mathcal{B}$ to use $\mathcal{A}$ (supposing he can distinguish between hops) to solve DDHP. Accordingly, we can bound their advantages based on the above. First, we give game hops as follows.

- **Game 0:** This is the original IND-CPA game for our HDRS.

1. The simulator $\mathcal{S}$ runs HDRS.Setup to get system parameters $\mathrm{param}_{\mathsf{HDRS}}^{\mathsf{ch}} = \langle G_2, p, g, H_2 \rangle$. Then, it runs HDRS.KeyGen to output (pk_R, sk_R). $\mathcal{S}$ relays $(\mathrm{param}_{\mathsf{HDRS}}^{\mathsf{ch}}, pk_R)$ to $\mathcal{A}$.

2. $\mathcal{A}$ randomly chooses two messages of the same length $m_0, m_1 \in \{0,1\}$ where $|m_0| = |m_1|$, and generates $sk_S = (x_0, x_1, x_2)$. $\mathcal{A}$ sends m_0, m_1 to $\mathcal{S}$. $\mathcal{S}$ flips a coin $coin \leftarrow \{0,1\}$ and generates $c_{coin} = $ HDRS.Signcrypt$(m_{coin}) = (c_{coin,0}, c_{coin,1}, c_{coin,2}, c_{coin,3})$ by:

$$\beta \xleftarrow{R} Z_p^*, \quad c_{coin,0} = g^\beta, \quad c_{coin,1} = y_{1,S}^\beta, \tag{5.14}$$

$$c_{coin,2} = g^{m_{coin}}(h \cdot y_{0,R})^\beta, \quad c_{coin,3} = y_{1,R}^{m_{coin}} \cdot (y_{2,R}^\beta)^{x_{1,S}}. \tag{5.15}$$

$\mathcal{S}$ sends c_{coin} to $\mathcal{A}$.

3. Finally, $\mathcal{A}$ outputs a guess by $coin' \in \{0,1\}$. If $coin = coin'$, $\mathcal{A}$ wins and $\mathcal{B}$ outputs 1; else, $\mathcal{B}$ outputs 0.

- **Game 1:** The same as Game 0 except that $\mathcal{S}$ changes $y_{0,R}^\beta$ with $R_0 \in G_2$ during step 2 while computing c_{coin}:

$$h = H_2(\mathsf{CID}, pk_R), \quad \beta \xleftarrow{R} Z_p^*, \quad \boxed{R_0} \xleftarrow{R} G_2, \tag{5.16}$$

$$c_{coin,0} = g^\beta, \quad c_{coin,1} = y_{1,S}^\beta, \tag{5.17}$$

$$c_{coin,2} = g^{m_{coin}} h^\beta \boxed{R_0}, \quad c_{coin,3} = y_{1,R}^{m_{coin}} \cdot (y_{2,R}^\beta)^{x_{1,S}}. \tag{5.18}$$

- **Game 2::** The same as Game 1 except that $\mathcal{S}$ changes $y_{2,R}^\beta$ with $R_1 \in G_2$ during step 2 while computing c_{coin}:

$$h = H_2(\mathsf{CID}, pk_R), \quad \beta \xleftarrow{R} Z_p^*, \quad R_0, \boxed{R_1} \xleftarrow{R} G_2, \tag{5.19}$$

$$c_{coin,0} = g^\beta, \quad c_{coin,1} = y_{1,S}^\beta, \tag{5.20}$$

$$c_{coin,2} = g^{m_{coin}} h^\beta R_0, \quad c_{coin,3} = y_{1,R}^{m_{coin}} \cdot \boxed{R_1}^{x_{1,S}}. \tag{5.21}$$

Each next hop in the above games only made a negligible change to the former one, i.e. the modification of parameters are beyond adversary $\mathcal{A}$'s view; otherwise, we can construct an algorithm $\mathcal{B}$ to solve DDHP by using $\mathcal{A}$ who can distinguish between **Game 0** and **Game 1**, or **Game 1** and **Game 2**. Take **Game 0** and **Game 1** as an example:

On given a DDHP instance $g, g^a, g^b, R \in G$, $\mathcal{B}$ decides by proceeding the following game.

1. $\mathcal{B}$ chooses $x_1, x_2 \overset{R}{\leftarrow} Z_p$, it sets $y_0 = g^a$ and computes $y_1 = g^{x_1}$ and $y_2 = g^{x_2}$. Next, $\mathcal{B}$ relays $pk_R = (y_0, y_1, y_2)$ and $\mathsf{param}_{\mathsf{LRRS}}$ to $\mathcal{A}$

2. $\mathcal{A}$ generates $sk_S = (x_0, x_1, x_2)$ as HDRS.KeyGen. Then, it samples $m_0, m_1 \overset{R}{\leftarrow} \{0,1\}$ and relays them to $\mathcal{B}$. $\mathcal{B}$ flips a coin $coin \leftarrow \{0,1\}$ and generates $c_{coin} = (c_{coin,0}, c_{coin,1}, Pc_{coin,2}, c_{coin,3})$ as:

$$h = H_2(\mathsf{CID}, pk_R), \quad \beta \overset{R}{\leftarrow} Z_p^*, \tag{5.22}$$

$$c_{coin,0} = g^\beta, \quad c_{coin,1} = y_{1,S}^\beta, \tag{5.23}$$

$$c_{coin,2} = g^{m_b} h^\beta Z, \quad c_{coin,3} = y_{1,R}^{m_{coin}} \cdot (y_{2,R}^\beta)^{x_{1,S}}. \tag{5.24}$$

3. Finally, $\mathcal{A}$ outputs his guess $coin' \in \{0,1\}$. If $coin' = coin$, $\mathcal{A}$ wins and $\mathcal{B}$ outputs 1; else, $\mathcal{B}$ outputs 0.

Denote E_i as the event of **Game i** that won by $\mathcal{A}$ (i.e. $1 \leftarrow \mathcal{S}$). If $Z = g^{ab}$ holds, this implies **Game 0**; else, if $Z \overset{R}{\leftarrow} G_2$, it implies **Game 1**. So, we can bound $\mathcal{B}$'s advantage in solving DDHP by $Adv_{\mathcal{B}}^{DDHP} = |Pr[E_0] - Pr[E_1]|$. Analogically, if $\mathcal{A}$ can distinguish between **Game 0** and **Game 1**, we can also bound $\mathcal{B}$'s advantage in solving DDHP by $Adv_{\mathcal{B}}^{DDHP} = |Pr[E_1] - Pr[E_2]|$.

In Game 2, $c_{coin} = \{c_{coin,0}, c_{coin,1}, c_{coin,2}, c_{coin,3}\}$ is generated following one-time pad (concretely, $c_{coin,0}$ and $c_{coin,1}$ are set as $\beta \overset{R}{\leftarrow} Z_p^*$, $c_{coin,2}$ is set as $R_0 \overset{R}{\leftarrow} G_2$, $c_{coin,3}$ is set as $R_1 \overset{R}{\leftarrow} G_2$), and therefore, $\mathcal{A}$'s view is beyond the random coin $coin$. Then, $\mathcal{A}$'s advantage in winning the Game 2 is negligible, i.e. $Pr[E_2] = \frac{1}{2}$. Analogically, from all above, we can bound the adversary $\mathcal{A}$'s advantage in breaking our IND-CPA security with $\mathcal{B}$ which solves the DDHP as: $Adv_{\mathcal{A}}^{IND-CPA} \leq 2 \cdot Adv_{\mathcal{B}}^{DDHP}$. Since $Adv_{\mathcal{B}}^{DDHP}$ is negligible, therefore, $Adv_{\mathcal{A}}^{IND-CPA}$ is negligible as well. Refer to [191] for more details.

5.6.5   Efficiency Analysis of HDRS

We evaluate the efficiency of our HDRS by complexity theory. We compare our HDRS with similar works in Table 5.2. For ease of comparison, we also show an overview of signcryption costs in Figure 5.9. As it is shown in Table 5.2, de-signcryption generally costs twice more complexities than the signcryption. Our HDRS ranks the 6th efficient scheme in the list. Specifically, assume each chosen group is equally large, i.e. $|G| = |G_1| = |G_2| = |G_T| = |Z_q^*|$. The communication cost of our HDRS is the smallest among all peer works. However, due to intractability to compute DLP, our decryption complexity is not measurable and assumed to be computationally exhausting. Refer readers to ref. [190] Section 7 for detailed analysis.

TABLE 5.2: Complexity of Signcryption Schemes

Scheme	Signcryption	De-Signcryption	Communication
PSIB [189]	$4T_e \approx 84T_m$	$2T_e + 2T_p + 2T_i \approx 227.6T_m$	$2|G_1| + 2|G_2| + |Z_q^*|$
AISI [185]	$4T_e \approx 84T_m$	$6T_p + T_i \approx 533.6T_m$	$4|G_1| + |G_2|$
IOSL [181]	$2T_{pm} + T_{mpm} + T_e + 2T_m \approx 96T_m$	$3T_{pm} + T_e + 2T_p \approx 261T_m$	$3|G| + n + 2|Z_q^*|$
EIOE [184]	$T_{mpm} + 3T_{pm} + 2T_m \approx 97T_m$	$T_p + T_{pm} + 2T_m + 2T_i \approx 122.6T_m$	$2|G| + n + 2|Z_q^*|$
OOIB [182]	$4T_{pm} + T_{mpm} + 3T_m \approx 120T_m$	$T_{mpm} + T_p \approx 116T_m$	$4|G| + n + 3|Z_p|$
HDRS [190]	$8T_e + 3T_m \approx 171T_m$	$T_e + T_m + T_i + T_{log} \geq 33.6T_m + T_{log}$	$4|G|$
FSIS [187]	$4T_e + T_p \approx 171T_m$	$6T_p + T_i \approx 533.6T_m$	$4|G_1| + 2|G_2|$
IBSS [188]	$6T_e + T_p \approx 213T_m$	$2T_e + 6T_p + T_i \approx 575.6T_m$	$4|G_1| + |Z_q^*|$
SISS [186]	$6T_e + T_p \approx 213T_m$	$2T_e + 6T_p + T_i \approx 575.6T_m$	$4|G_1| + |G_2| + |Z_q^*|$
IBOO [183]	$6T_{mpm} + 2T_e + 2T_m \approx 218T_m$	$2T_p + 5T_{mpm} \approx 319T_m$	$|G_T| + 5|G_T| + n + 2|Z_p^*|$

Listed by signcryption cost from low to high. Each scheme is named by the abbreviation of title.

5.7 Methods to Construct an IND-CCA2 Secure Encryption Scheme

In this section, we enumerate several generic methods to either achieve IND-CCA2 security or transform any insecure encryption into an IND-CCA2 one.

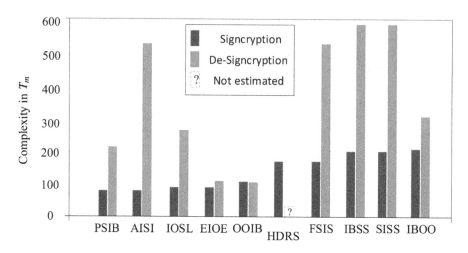

FIGURE 5.9: Overview of Signcryption Schemes

5.7.1   Generic Methods to Achieve IND-CCA2

So far, IND-CCA2 security is widely accepted as the ideal and practical security level for any encryption schemes to be applied in various applications (including blockchain). In this section, we will discuss several methods to achieve IND-CCA2 security. IND-CCA2's difference with IND-CCA1 is that the adversary is allowed to gain useful information after receiving the challenged ciphertext c_b but before launching the attack. .

Generally, there are several approaches to achieve an IND-CCA2 security for encryption schemes:

1. Non-Interactive Zero-Knowledge Proof (NIZK) Type.

2. Random Oracle Model (ROM) Type.

3. Universal Hash Function Type (UHF).

4. Conversion of Identity-Based Encryption.

5. Hybrid Encryption (HY).

6. Diffie-Hellman Integrated Encryption Scheme (DHIES) [192]: An extension of ElGamal scheme [94].

5.7.2   Background of IND-CCA2

We give an overview of the background of IND-CCA2 in Figure 5.10.

We classify the history of IND-CCA2 into three stages with a brief description as follows.

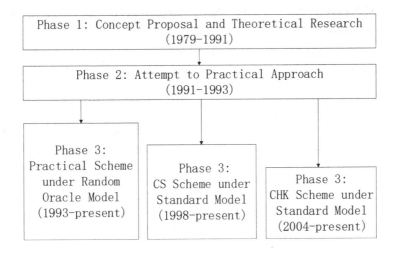

FIGURE 5.10: Development of Provable Security for PKE Scheme

1. **Phase 1:** Conceptual and Theoretical Research

    1979–1984. Rabin proposed the first formalized provable secure PKE scheme [193].

    1984–1990. Goldwasser and Micali first defined IND-CPA [194].

    1990. Naor and Yung first defined IND-CCA [153].

    1991. Rackoff and Simon first defined IND-CCA2 [6].

2. **Phase 2:** Practical Scheme

    1991. Damgard proposed two efficient PKE schemes.

    1992. Zheng and Seberry proposed three efficient PKE schemes [195]. They gave the main idea for IND-CCA2 and prototype for 'signcryption' [170].

    1993. Lim and Lee presented an efficient scheme.

3. **Phase 3:** Three Types of Provably Secure PKE

    Random Oracle Model (ROM) based Schemes.

    Cramer and Shoup's Scheme [169,196] (CS series) under Standard Model.

    Canetti, Halevi, and Katz's scheme [197] (CHK) under Standard Model.

5.7.3   IND-CCA2   from   Non-Interactive   Zero-Knowledge (NIZK)

Fiat, Fiege, and Shamir [198], for the first time, defined zero-knowledge proof. Then, Galil, Haber, and Yung constructed an interactive PKE provably secure under IND-CCA model out of a trapdoor function [199]. Blum, Feldman, and Micali [200] then proposed non-interactive Zero-knowledge, which then formally defined by Rackoff and Simon [6].

In NIZK, two separate keys were used to encrypt the message, and the proof was made to show these two keys are implemented to the same message for encryption. The resulting encryption scheme, uses a generic reduction from an instance of encryption scheme to an instance of some NP-complete problem [200].

There are currently no known efficient NIZK proof systems even under specific assumptions and particular cryptosystems of interest.

5.7.4   IND-CCA2 from Random Oracle Model

The notion "ROM" was first proposed by Goldreich et al. in [201,202]. Bellare and Rogaway formally defined the security proof in ROM [26]. In 1993, Bellare and Rogaway [203] proposed optimal asymmetric encryption padding (OAEP) from one-way trapdoor permutation IND-CCA1. This work was immediately followed by OAEP+ [203], and OAEP-RSA [204] against CCA2. In 1999, Fujisaki and Okamoto [205] proposed general constructions from IND-CPA to IND-CCA2. Then, in 2000, Pointcheval [206] proposed generic conversion from an arbitrary one-way trapdoor function to IND-CCA2. In 2001, Okamoto and Pointcheval [207] proposed rapid enhanced-security asymmetric cryptosystem transform (REACT) as a new conversion which applies to any weakly secure cryptosystem to realise practical security. Fujisaki and Okamoto [205] proposed the first generic method that transforms any probabilistic trapdoor one-way function to an IND-CCA2 scheme in the random oracle model, provided that the trapdoor one-way function is semantically secure (IND-CPA).

To note, an asymmetric encryption scheme is one-way if an adversary cannot entirely decrypt the encryption of a random plaintext, but this notion is too weak to protect the privacy of messages because it only guarantees that $w(\log k)$ bits of the plaintext is infeasible to determine. It is worth noticing that Fujisaki and Okamoto eliminated the IND-CPA's requirement for trapdoor one-way function in [168]; this is a typical attempt to achieve IND-CCA security. However, that solution only yields IND-CCA1 security.

5.7.5   IND-CCA2 from (UHF)

We give some background information for UHF. Briefly, the universal hash proof (UHF) system is a special kind of non-interactive zero-knowledge proof system for a language. It can be used as a primitive to construct a secure cryptographic IND-CCA2 scheme. Cramer and Shoup first formalized this notion in ref. [196]. In 1998, Cramer and Shoup [169] presented the first truly practical IND-CCA secure under the standard model based on DDH Assumption. In 2002, a general construction for the above was proposed in ref. [196]. The generic framework was extended by using a new cryptographic primitive called universal hash proof system. In 2003, a hybrid encryption was formally discussed by Shoup [208] to encrypt arbitrary messages, in which key encapsulation mechanism (KEM) and data encapsulation mechanism (DEM) were proposed.

Before diving into UHF cases, we must know the notion of one-way trapdoor function (OWTP) and the standard model. A one-way trapdoor function means easy to compute in one direction yet tricky in the opposite direction without the trapdoor. A formal proof of security for a public key cryptosystem relying solely on the intractability of the cryptosystem's underlying OWTP is said to be a proof under standard intractability assumption. Such a proof establishes security in the real world; it proves that a cryptosystem cannot be broken without breaking the underlying intractability assumptions.

Standard Model is a basic model of reasonable assumption, such as target collision-resistant hash function (also known as a universal one-way hash function, UWF). Although the standard model is more acceptable and dependable on security, such rigorous proof often results in an impractical scheme or nothing.

5.7.5.1 Universal Hash Function

In CS scheme [169,196], a family of hash functions is said to be collision-resistant if upon drawing a function H at random from the family, it is infeasible for an adversary to find two different inputs x and y such that $H(x) = H(y)$.

A weaker notion [209] is the universal one-way family of hash functions. Here, it should be infeasible for an adversary to choose an input x, compute a random hash function H, and then find a different input y such that $H(x) = H(y)$. Also, it can be called a target collision-resistant hash function. For algebraic definition of UHF, refer to ref. [196].

5.7.5.2 UHF Case: CS Scheme and Its Variants

Previously, we explicitly analyzed Cramer and Shoup's scheme in Section 5.5. As the reader shall find out in Section 5.5.2, the given proof relies on the assumption of a universal one-way family. Also, as concluded by Cramer and Shoup in the literature [169], the use of universal one-way hash function suffices to prove the security of the CS scheme.

Following the basic CS scheme [169], Cramer and Shoup generalized their construction by considering an algebraic primitive they call universal hash proof systems [196]. Notably, they showed that this framework yields not only the original DDH-based Cramer Shoup's scheme but also encryption schemes based on quadratic residuosity (QR) and on Paillier's Decision Composite

Residuosity (DCR) assumption [210]. In [196], their construction only uses the universal hash proof system as a primitive; no other primitives are required.

5.7.6   IND-CCA2 from Hybrid Encryption (HY) and Use Case

Since traditional public-key encryption schemes limit the message domain in a certain group, it was unavailable to deal with the message of arbitrary length. To tackle, hybrid encryption (HY) was proposed; it used symmetric encryption for the message and asymmetric encryption for encrypting keys. In 2003, Cramer and Shoup gave a formal design of HY, namely, key encapsulation mechanism (KEM) and data encapsulation mechanism (DEM) [208]. Later, Kurosawa and Desmedt [211] lessened the requirement of KEM and DEM.

Here, we instantiate HY's use by exemplifying Fujisaki and Okamoto's proposal [168] as a use case. We denote it as Fujisaki and Okamoto (FO) scheme. As given in Figure 5.11, we mainly describe the encryption and decryption algorithms. Briefly, Fujisaki and Okamoto show a generic conversion from weak asymmetric and symmetric encryption schemes into an IND-CCA2 secure asymmetric encryption in the random oracle model. As can be observed from Figure 5.11, the encrypted message consists of ciphertexts from the public key encryption and symmetric encryption, respectively. The decryption proceeds by recovering encryption key from ciphertext under the public key encryption and using the derived key to recover plaintext from ciphertext under symmetric key encryption. Refer to ref. [168] for more details.

5.7.7   IND-CCA2 from ElGamal and Its Extensions

ElGamal [94] is a classical PKE scheme but not an IND-CCA2 secure one. However, following the above generic methods or other techniques, one can strengthen ElGamal and its variants to achieve IND-CCA2. Many encryptions are converted from ElGamal, including CS scheme [169] and KD04's scheme [211] based on DDH assumption, Kiltz07's scheme [212] based on GHDH, and DHIES's scheme [192] based on ODH scheme.

We give a chart to compare the ciphertext of the ElGamal scheme with other IND-CCA2 secure schemes in Figure 5.3. The reader shall observe the insecurity of the ElGamal scheme. We consider the exploration of this insecurity as a learning process. Readers are referred to ref. [213] for more details. As shown in Figure 5.3, CS and KD04 are based on weaker DDH assumption and sacrifice efficiency. Kiltz07 scheme's security is based on stronger GHDH assumption. However, DHIES is based on relatively stronger assumption ODH than former ones with the best efficiency. We give a comparison of variants of ElGamal scheme with IND-CCA in Table 5.3.

Exemplify HY: Fujisaki and Okamoto's Scheme

σ denotes a random string; ε_{pk}^{asym} denotes an asymmetric encryption.

ε_a^{sym} denotes a symmetric encryption; G and H are hash functions.

$COIN$ denotes a randomness space.

KeyGen

1. $(pk, sk) \leftarrow \mathcal{K}^{hy}(1^k)$.

Encryption

2. Compute $h = H(\sigma, c)$ and $e = \varepsilon_{pk}^{asy}(\sigma, h)$.
3. Compute $c = \varepsilon_a^{sy}(m)$ and $a = G(\sigma)$
4. $\varepsilon_{pk}^{hy}(m, \sigma) = e \parallel c$

Decryption

5. Compute $\hat{\sigma} = \mathcal{D}_{sk}^{asy}(e)$.

6. If $\hat{\sigma} \in COIN^{hy}$, compute $\hat{a} = G(\hat{\sigma})$; otherwise, set $\mathcal{D}(e \parallel c) = \varepsilon$ and go to step 9.
7. Set $\hat{h} = H(\hat{\sigma}, c)$.

8. If $e = \varepsilon_{pk}^{say}(\hat{\sigma}, \hat{e})$, set $\mathcal{D}_{sk}^{hy}(e \parallel c) = \mathcal{D}_{\hat{a}}^{sy}(c)$; otherwise, set $\mathcal{D}_{sk}^{hy}(e \parallel c) = \varepsilon$.
9. Return $m = \mathcal{D}_{sk}^{hy}(e \parallel c)$.

FIGURE 5.11: The sketch of FO scheme

TABLE 5.3: Comparison of ElGamal Scheme with Other IND-CCA2 Schemes

Version	Ciphertext	Assumption
ElGamal [94]	(g^k, y^k)	CDH
CS [169]	$(g_1^r, g_2^r, h^r m, c^r d^{ra})$	DDH
KD04 [211]	$(g_1^r, g_2^r, e = SKE.Enc_{k_1}(m),$ $MAC_{k_2}(e)), k_1 \parallel k_2 = H(c^r d^{ra})$	DDH
Kiltz07 [212]	$(g^r, u^{ra} v^r), k = H(u^r)$	GHDH
DHIES [192]	$(g^r, e = SKE.Enc_{k_1}(m),$ $MAC_{k_2}(m)), k_1 \parallel k_2 = H(h^r)$	ODH

5.8 Chapter Summary

This chapter leverages the rigorous security model and proof techniques to achieve the practical design and PKE scheme analysis. It begins with an overview of existing PKE schemes. Then, it systematically analyzes the design and security of two representative IND-CCA2 secure PKE schemes: Identity-Based Encryption (IBE) and Cramer and Shoup's encryption (CS) schemes. To exemplify the proof technique, it also gives a sample of our IND-CPA secure encryption scheme. Next, it summarizes generic methods to achieve IND-CCA2 security. After reading this chapter, the reader should be able to: (1) know how to define public-key encryption scheme and its security model; (2) understand the instantiated proving techniques by practising on the given examples; (3) sketch the basic design of IBE and CS schemes and evaluate its security via one of the proving techniques mentioned in this chapter; (4) and enumerate and explain generic methodologies to achieve IND-CCA2 security. To summarize, it is challenging to achieve both IND-CCA2 security and practical design of an algorithm for blockchain. The reasons are that: (1) Achieving IND-CCA2 relies on either impractical tools, strong assumptions, trusted entities, or setup phase, (2) Blockchain is generally decentralized and does not support computing from any centralized or trusted entities, (3) Programming security reduction for practical PKE scheme requires highly empirical knowledge and sophisticated skills.

Chapter 6

Public-Key Hash Function for Blockchain

6.1 Chapter Introduction

Public key hash (PKH) scheme is an easy starting point for a beginner, which has evolved from the hash function but involves the public and the private keys for computation. Similar to other public key primitives, the private key is used to compute the secrecy like a trapdoor, while the public key is used for the verification. Recent advances in blockchain have witnessed uses of PKH to build the redactable blockchain where block hash and transaction are supposed to be redacted without impairing consistency. To study PKH systematically, this chapter dives deeply into the design and analysis of current PKH schemes. Concretely, this chapter first reviews the background of PKH. Then, it instantiates by various PKH schemes: ACH, HCCH, PCH, TUCH and RCH which are mostly from our published works.

After reading this chapter, you will:

- Know how to define the public key hash scheme and its security model.

- Grasp the proving techniques by learning and practicing the instantiated examples.

- Give formal security analysis for any instantiated examples of PKH schemes in this chapter.

6.2 Overview of PKH

Hash function is generally used and known for producing digest with the fixed length of the arbitrary message. Conventionally, the hash function is used by a symmetric manner and does not require to be part of the public key infrastructure (PKI) [3]. However, the involvement of PKI leads to diverse use and a new notion called public-key hash function. Inspired by chameleon hash (and view it as a special variant of public key hash scheme), we formalize the notion of public key hash (PKH) in this chapter. We interchangeably use public key hash (PKH) and chameleon hash (CH) in this book, as they both lead to the same notion we seek to formalize in this chapter. We begin by reviewing some background information.

Krawczyk and Rabin [85] first proposed chameleon hash as a tool to achieve the non-interactive version of the undeniable signature, i.e. chameleon signature. Briefly, chameleon hash is a trapdoor one-way hash function where it is hard to compute hash collision without private key (also known as the trapdoor key). In comparison with a general hash function (say $H(\cdot)$), chameleon hash ($CH(\cdot)$) involves public key with hashing computation which allows the holder of the corresponding private key to compute its hash collision efficiently, i.e. finding (m', r') such that $CH(m, r) = CH(m', r')$ where r' denotes a new randomness. Chameleon hash is characterized by several features:

1. One-wayness;

2. Collision-resistance;

3. Computationally efficient;

4. Uniformly distributed.

A similar notion is the trapdoor commitment [214] where a user commits his knowledge to a single value but reveal it later for the verification. Specifically, there is a trapdoor for users to change the committed knowledge without impairing the revelation performs later. For ease of understanding, the chameleon hash can be viewed as a simple version of trapdoor commitment.

6.2.1   PKH in Blockchain

Blockchain [1] is a convenient and effective trust-layer for the above concepts. When applying blockchain as a trust-layer, security and privacy rules are applied. However, as recently reported by Ali et al. [42], the newly launched blockchain suffer from powerful attacks (such as 51% attack [1]) due to small the computing pool. As we will explicitly explain in Section 8.5.1, the chameleon hash function can be utilized as a countermeasure against the inflexibility and vulnerabilities found in traditional immutable blockchains. To explain, malicious users can easily manipulate the blockchain with considerable, but not unimaginably-large computing resources. To address this problem, Ateniese et al. [215] introduced the notion of redactable blockchain by adopting the chameleon hash to compute bock hash instead of SHA-256. This allows removing illegal and malicious contents from the blockchain without causing inconsistency of the block hash. Related works are summarized and discussed in Section 8.5.1.

Although there are no convincing use cases to state, the studies of the chameleon hash are gaining massive attention in both academia [216] and industry [217].

6.2.2 Introduction to PKH

Briefly, the chameleon hash is a trapdoor one-way hash function where finding a collision is challenging without the trapdoor key. The first idea was proposed by Krawczyk et al. [218] where the notion of chameleon signature (CS) was also revealed. Later on, the key-exposure problem is identified by Ateniese et al. [219], which captures the leakage of the trapdoor key in any given hash collisions. Since then, several schemes with key-exposure freeness were proposed, such as [86,148,220]. We review chameleon hash schemes in Table 6.1, the CH schemes are compared based on the involvements of ID-based infrastructure, pairing computations, collision-resistance and intractability assumptions. As shown in Table 6.1, most of the listed schemes are not based on ID-based infrastructure and pairing computations.

TABLE 6.1: Overview of Chameleon Hash Works

Scheme	Identity -based	Pairing -based	Collision -Resistance	Assumptions
ICH [219]	Yes	No	Yes	q-SDHP
KEF [148]	No	No	Yes	CDHP
CHE [86]	No	No	Yes	RSA and DLP
CHD [220]	No	No	Yes	RSA
ACH [221]	No	Yes	Yes	CDHP
CHW [149]	No	No	Yes	CDHP
CIK [222]	No	No	Yes	Factoring
ACC [223]	No	No	Yes	Factoring

Each work is named after the abbreviation of article title.

6.2.3 Defining PKH

A PKH scheme consists of the following four algorithms.

$\mathsf{SysGen}_{\mathsf{PKH}}$: This algorithm takes as input a security parameter λ. It returns the system parameters SP.

$\mathsf{KeyGen}_{\mathsf{PKH}}$: This algorithm takes as input SP. It returns a public/secret key pair (pk, sk).

$\mathsf{Hash}_{\mathsf{PKH}}$: This algorithm takes as input a message m, pk, and a randomness r. It returns a hash value $\hbar$.

$\mathsf{Verify}_{\mathsf{PKH}}$: This algorithm takes as input a tuple $(m, r, \hbar)$. It returns 0 or 1.

$\mathsf{Forge}_{\mathsf{PKH}}$: This algorithm takes as input a tuple $(m, r, \hbar)$, and a new message m', it returns a new randomness r'.

The validation of a PKH scheme generally consists of correctness and security analysis based on security requirement. We generalize them as follows:

Correctness. Given any $(pk, m, r, \hbar)$, if $\hbar$ is a valid hash of m under randomness r, the verification algorithm will return "1". It is not necessarily needed to define a verification algorithm in concrete design since the user can redo the hashing scheme again to check the validity of hash.

Security. Without the secret key sk, it is hard for any probabilistic polynomial time (PPT) adversary to forge a valid hash collision (m', r') for a given tuple $(m, r, \hbar)$ that satisfies the correctness.

In the security model of chameleon hash, the security is modeled by a game played among a challenger and an adversary, where during the interaction between them, the challenger generates a chameleon hash, and the adversary tries to break the scheme. There are two significant security models for PHK: Indistinguishability (IND) and Collision-Resistance (CR). Specifically, in a game defined for CR security, the challenger first generates a key pair (pk, sk), sends the public key pk to the adversary, and keeps the secret key. The adversary can then make hash queries on any messages adaptively chosen by the adversary itself. Finally, the adversary returns a forged hash collision of a new message that has not been queried.

The security model of collision-resistance (ColRes) can be described as follows.

Setup. Let SP be the system parameters. The challenger runs the key generation algorithm to generate a key pair (pk, sk) and sends pk to the adversary. The challenger keeps sk to respond to signature queries from the adversary.

Query. The adversary makes hash queries on messages that are adaptively chosen by the adversary itself. For a hash query on the message m_i, the challenger runs the signing algorithm to compute σ_{m_i} and then sends it to the adversary.

Forgery. The adversary returns a forged collision $(m^*, r^*, \hbar^*)$ and $(m^{*'}, r^{*'}, \hbar^*)$ and wins the game if

- $\hbar^*$ is a valid hash of the tuples $(m^*, r^*, \hbar^*)$ and $(m^{*'}, r^{*'}, \hbar^*)$,

- A hash of m^* has not been queried in the query phase.

The advantage ϵ of winning the game is the probability of returning a valid forged signature. Similarly, one can define the indistinguishability of each hash and randomness. We omit the detail. Refer to ref. [86] for more details.

6.3 Case Analysis: ACH

In this section, we review the construction and security analysis of accountable chameleon hash (ACH) scheme proposed by Lai et al. [221].

6.3.1 Construction of ACH

We give a sketch of ACH construction in Figure 6.1. Detailed construction can be found in ref. [221]. The sketch suffices for the reader to follow security analysis.

Accountable Chameleon Hash (ACH) Scheme
System parameter: param$_{\text{ACH}}$ = $\{p, G, G_T, \hat{e}, g, H\}$; Public key: $pk_i = g^{x_i}$; Private key: $sk_i = x_i$.

Hash	Forge
1. Given $pk_i, pk_j, m \in Z_p^*$ and TID, pick a random $R \in G$ 2. Compute $\hbar = \hat{e}(R, g) \cdot \hat{e}(H(\text{TID}^m), pk_i \cdot pk_j)$.	3. Given $sk_i = x_i$ and TID, compute: $td_{i,\text{TID}} = H(\text{TID})^{x_i}$ 4. Given sk_j, $td_{i,\text{TID}}$, R and m', compute $R' = R \cdot (H(\text{TID})^{sk_j} \cdot td_{i,\text{TID}})^{m-m'}$.

FIGURE 6.1: Sketch of ACH Scheme

6.3.2 Security Analysis of ACH

Proof Overview 6.1 Security of ACH mainly consists of collision resistance, existential forgery, and forgery indistinguishability. Transitions prove the former based on the failure event. The latter is proved by deduction and case analysis. Based on the previous studies in BLS, the reader find the below easy to follow. The analysis begins with a game constructed between adversaries $\mathcal{A}$ and $\mathcal{B}$. $\mathcal{B}$ seeks to exploit $\mathcal{A}$'s attack in order to solve the CDH problem. The bound of $\mathcal{B}$'s probabilities in solving CDHP can be derived from computing the probabilities of the abortion. This follows the same pattern we previously analyzed for BLS and IBE in Sections 4.4.3 and 5.4.4.

To note, Lai et al. [221] prove indistinguishability by merely analysing the distribution domain. Although this should suffice for analysis, it is recommended that the reader programme a formal analysis based on the security model. As this will be more convincing and inspiring for cryptographic researches. We simplify Theorem 6.1 and 6.2 for ease of understanding.

Theorem 6.1 *6.1 ACH is secure against collision of adversaries, assuming that the CDH assumption holds.*

Proof of Theorem 6.1. Suppose there exists an adversary $\mathcal{A}$ against the collision resistance of the above ACH scheme. One can construct a PPT $\mathcal{B}$ that uses $\mathcal{A}$ to solve the CDHP efficiently. $\mathcal{B}$ is given a tuple (g, g^a, g^b), and

$\mathcal{B}$'s goal is to compute g^{ab}. $\mathcal{B}$ runs $\mathcal{A}$ to derive an answer to CDHP instance. We omit the details here. Refer to Section 4.4.2 or ref. [221] for more details.

Tip 6.1 Above completes the proof of Theorem 6.1. One should find it easy to follow based on the empirical knowledge. Next is the proof of indistinguishability. This is not a formal analysis since it does not follow a standard security model as we previously noted. More formal analysis for indistinguishability of CH will be given in Sections 6.4.3 and 6.6.3.

Theorem 6.2 *The above construction of ACH is forgery indistinguishable.*

Proof of Theorem 6.2. The ACH scheme is said to be forgery indistinguishable the following distribution ensembles are computationally indistinguishable:

$$\mathcal{D}_{\mathsf{Forge}} = \{(m', \hat{R}, \hbar) | R \xleftarrow{\$} \mathcal{R}, \hbar \leftarrow Hash(param, pk_i, pk_j, \mathsf{TID}, m, R), td_{i,\mathsf{TID}} \leftarrow$$

$$\mathsf{Trapdoor}(sk_i, \mathsf{TID}), \hat{R} \leftarrow \mathsf{Forge}(sk_j, \mathsf{TID}, pk_i, pk_j, \mathsf{TID}, m, R, \hbar, m')\}_{\lambda, param, pk_i, pk_j, \mathsf{TID}}$$

$$\mathcal{D}_{\mathsf{Hash}} = \{(m', \hat{R}', \hbar) | R' \xleftarrow{\$} \mathcal{R}, \hbar' \leftarrow Hash(param, pk_i, pk_j, \mathsf{TID}, m, R), td_{i,\mathsf{TID}} \leftarrow$$

$$\mathsf{Trapdoor}(sk_i, \mathsf{TID}), \hat{R} \leftarrow \mathsf{Forge}(sk_j, \mathsf{TID}, pk_i, pk_j, \mathsf{TID}, m', R')\}_{\lambda, param, pk_i, pk_j, \mathsf{TID}}$$

One can easily conclude the above two distributions' computational indistinguishability by enumerating each parameter carefully and giving an deduction accordingly. Since this does not involve skilled proof technique, we omit details here.

6.4   Case Analysis: HCCH

This section reviews our original work of homomorphic collision-resistant chameleon hash (HCCH) [190]. We mainly give the construction and the security analysis based on the security requirements.

6.4.1 Construction of HCCH

We give a sketch of construction for HCCH in Figure 6.2. We also give a simplified algorithm. This shall suffice the reader to follow security analysis.

Homomorphic Collision-resistant Chameleon Hash (HCCH) Scheme		
System parameter: $param_{HCCH} = \{G_1, p, g, H_1\}$; Public key: $hk = y = g^x$; Private key: $tk = x \xleftarrow{R} Z_p^*$		
Hash	Verify	Forge
1. Given CID and $M \in \{0,1\}^*$, pick $\alpha \xleftarrow{R} Z^*$.	5. Given $(M, \hbar, r)$, compute $e = H_1(CID, y)$.	9. Given $M' \in \{0,1\}^*$ and $(M, \hbar, r)$, check validity of $\hbar$.
2. Compute $e = H_1(CID, y)$ and $h = g^e$.	6. Check whether $< g, g^\alpha, y, y^\alpha >$ is a DH tuple.	10. Compute $e = H_1(CID, y)$ and $r' = g^{(H_1(M) - H_1(M'))(x+e)^{-1}}, y^\alpha \cdot$ $y^{(H_1(M) - H_1(M'))(x+e)^{-1}}$.
3. Compute $r = (g^\alpha, y^\alpha)$.	7. Check whether $< g, g^\alpha, h \cdot y, \frac{\hbar}{g^{H_1(M)}} >$ is a DH tuple.	11. Check validity of r'.
4. Compute $\hbar = g^{H_1(M)}(h \cdot y)^\alpha$.	8. If yes, output 1; otherwise, output 0.	12. If yes, output 1; otherwise, output 0.

FIGURE 6.2: Sketch of HCCH Scheme

HCCH.Setup$(\lambda) \rightarrow (param_{TUCH})$
HCCH.KeyGen$(param_{TUCH}) \rightarrow (tk, hk)$
HCCH.Hash$(hk, CID, M, \alpha) \rightarrow (\hbar, r)$. Parse chameleon randomness as $r = (g^\alpha, y^\alpha)$ and chameleon hash as $\hbar = g^{H_1(M)}(h \cdot y)^\alpha$.
HCCH.Verify$(hk, CID, (M, \hbar, r)) \rightarrow (0 \text{ or } 1)$
HCCH.Forge$(tk, CID, M', (M, \hbar, r)) \rightarrow (r' \text{ or } \perp)$
HCCH.Int$(hk, (H_1(M_1), \mathcal{H}_1, r_1), \cdots, (H_1(M_n), \mathcal{H}_n, r_n), CID) \rightarrow ((\tilde{M}, \tilde{\mathcal{H}}, \tilde{r}) \text{ or } \perp)$

6.4.2 Security Requirement of HCCH

We capture the security requirement of HCCH by following two experiments: $IND_\mathcal{A}^{HCCH}$ and $PColRes_\mathcal{A}^{HCCH}$). The following definitions suffice to give an overview of the game where the security reduction is supposed to be programed on.

Experiment: $IND_\mathcal{A}^{HCCH}(\lambda)$
  $param_{TUCH}^{ch} \leftarrow$ HCCH.Setup(λ)
  $(hk_{ch}, tk_{ch}) \leftarrow$ HCCH.KeyGen$(param_{TUCH}^{ch})$
  $b \leftarrow \{0, 1\}$
  $a \leftarrow \mathcal{A}^{\mathcal{O}^{Hash\&Forge}(tk_{ch}, \cdots, a), HCCH.Forge(tk_{ch}, \cdots)}(hk_{ch})$
    where oracle $\mathcal{O}^{Hash\&Forge}$ on input
    (tk_{ch}, CID, M, M', b):
      Set $(\hbar, r) \leftarrow$ HCCH.Hash$(hk_{ch}, CID, M, \alpha)$
      Set $(\mathcal{H}', r') \leftarrow$ HCCH.Hash$(hk_{ch}, CID, M', \alpha')$
      Set $(r'') \leftarrow$ HCCH.Forge$(tk_{ch}, CID, M, (M', \hbar, r'))$

If HCCH.Verify(hk_{ch}, CID, M', $\mathcal{H}'$, r') = $\perp$
$\quad \lor r'' = \perp$, return $\perp$
$\quad$ If $b = 0$, return $(\hbar, r)$
$\quad$ If $b = 1$, return $(\hbar', r'')$
return 1, if $a = b$;
else, return 0.

Experiment: $PColRes_{\mathcal{A}}^{HCCH}(\lambda)$

$\quad param_{TUCH}^{ch} \leftarrow$ HCCH.Setup(λ)
$\quad (tk_{ch}, hk_{ch}) \leftarrow$ HCCH.KeyGen($param_{TUCH}^{ch}$)
$\quad (CID^*, M^*, r^*, M^{**}, r^{**}, \hbar^*) \leftarrow \mathcal{A}^{Forge'(tk_{ch},\cdots)}(hk_{ch})$
$\quad\quad$ where oracle $Forge'$ on input $(tk_{ch}, CID, M, M', r, \hbar)$:
$\quad\quad$ Return $\perp$ if $0 \leftarrow$ HCCH.Verify(hk_{ch}, CID, M, $\hbar$, r)
$\quad\quad (r') \leftarrow$ HCCH.Forge(tk_{ch}, CID, M', $(M, \hbar, r)$)
$\quad\quad$ If $r' = \perp$, return $\perp$
$\quad\quad \mathcal{Q} \leftarrow \mathcal{Q} \cup \{CID\}$ where $\mathcal{Q}$ denotes a list of query history
$\quad\quad$ Return r'
$\quad$ Output 1 if $1 \leftarrow$ HCCH.Verify(hk_{ch}, CID^*, $(M^*, \hbar^*, r^*)$) = 1 $\land$
$\quad 1 \leftarrow$ HCCH.Verify(hk_{ch}, CID^*, $M^{**}, \hbar^*, r^{**}$)
$\quad \land CID^* \notin \mathcal{Q} \land M^* \neq M^{**}$
else, return 0, if $a = b$.

6.4.3   Security Analysis of HCCH

Proof Overview 6.2 Security of HCCH mainly consists of indistinguishability and public collision resistance. The former is proved by transitions based on indistinguishability. The latter is proved by deduction and proof by the theorem. As one shall discover, the proof of collision resistance for CH is similar to the proof of existential unforgeability of a public signature (given in Section 4). The proof of public collision resistance for ACH borrows the idea from Boneh, and Boyen's short signature scheme [32]. Based on previous studies, the reader shall find it easy to follow the proofs below. Although we did not exemplify short signature scheme [32], the reader refers to ref. [32] for more details.

6.4.3.1   Proof of Indistinguishability

We prove by game hopping as follows:

- **Game 0:** This is the original indistinguishability game where $b = 0$.

- **Game 1:** The same as Game 0 except hashing directly to derive $\hbar$.

- **Game 2::** The same as Game 1 except hashing directly to derive $\hbar'$.

Denote E_i as the event of **Game i** won by $\mathcal{A}$ (i.e. $1 \leftarrow IND_{\mathcal{A}}^{\mathsf{HCCH}}(\lambda)$). Set **Game 0** as the original game defined in $IND_{\mathcal{A}}^{\mathsf{HCCH}}(\lambda)$. The advantage of $\mathcal{A}$ in winning **Game 0** is $Adv_{\mathcal{A}}^{IND} = |Pr[E_0] - \frac{1}{2}|$.

Transition from Game 0 to game 1: This hop only modifies the view of adversary $\mathcal{A}$ negligibly due to the indistinguishability of our HCCH. Otherwise, if $\mathcal{A}$ can distinguish this hop, an adversary $\mathcal{B}$ can be constructed to break the indistinguishability of HCCH. Concretely, $\mathcal{B}$ replaces hk_{ch0} with hk_{ch1} and queries the oracle $\mathcal{O}^{\mathsf{Hash\&Forge}}$ to derive $\hbar$. Then, $\mathcal{B}$ relays the output to $\mathcal{A}$. Due to the indistinguishability of our HCCH, we have $|Pr[E_0] - Pr[E_1]| \leq \nu(\lambda)$.

Transition from Game 1 to game 2: This hop only modifies the view of adversary $\mathcal{A}$ negligibly due to the indistinguishability of our HCCH. Otherwise, if $\mathcal{A}$ can distinguish this hop, an adversary $\mathcal{B}$ can be constructed to break the indistinguishability of HCCH. Concretely, $\mathcal{B}$ replaces hk_{ch1} with hk_{ch2} and applies the oracle $\mathcal{O}^{\mathsf{Hash\&Forge}}$ to derive $\hbar'$. Then, $\mathcal{B}$ relays the output to $\mathcal{A}$. Analogically, we have $|Pr[E_1] - Pr[E_2]| \leq \nu(\lambda)$.

Last, for $b = 1$ we have $Pr[E_2] = \frac{1}{2}$.

Based on the above, we can deduce that $Adv_{\mathcal{A}}^{IND} \leq Adv_{\mathcal{B}}^{IND} \leq \nu(\lambda)$. Since each hop modifies the view slightly and this change is beyond the adversary $\mathcal{A}$'s view, our HCCH is indistinguishable.

6.4.3.2 Proof of Public Collision-Resistance

Suppose $\mathcal{A}$ is an efficient adversary who breaks the indistinguishability of our HCCH, we briefly show how to construct a probabilistic polynomial time (PPT) algorithm $\mathcal{B}$ to solve q-SDHP [32]. On given a q-SDHP instance $(g, g^x, \cdots, g^{x^q})$, we denote it as $(A_0, A_1, \cdots, A_q)$ where $A_i = g^{x_i} \in G_1$ for $i = 1, \cdots, q$ and $A_0 = g$. Here, $x \in Z_p^*$ is unknown. In order to derive an answer $(c, g^{\frac{1}{(x+c)}})$ for some $c \in Z_q^*$ (which can either be designated [32] or not, for ease of analysis, we do not designate it). $\mathcal{B}$ interacts with $\mathcal{A}$ to derive an answer for q-SDHP as below:

Query: Adversary $\mathcal{A}$ issues q_s distinct queries $\{\mathsf{CID}_i, M_i', (M_i, \hbar, r_i)\}_{i \in [1, q_s]}$ under same hash key $hk = y$. Assume $q_s = q - 1$.

Response: For each M_i where $1 \leq i \leq q_s$, $\mathcal{B}$ generates the corresponding response as follows: Set polynomial $f(z) = \prod_{i=1}^{q_s}(z + e_i) = \sum_{i=0}^{q_s} a_i z^i$ where $a_0, \cdots, a_{q_s}$ are coefficients of polynomial $f(z)$, $e_i = H_1(\mathsf{CID}_i, y)$, and $hk = y$ is the hash key. Define:

$$g' = \prod_{i=0}^{q_s}(A_i)^{a_i} = g^{f(z)} \text{ and } \tilde{h} = \prod_{i=1}^{q_s}(A_i)^{a_{i-1}} = g^{zf(z)} = g'^z. \tag{6.1}$$

Next, we define polynomial $f_i(z) = f(z)/(z + e_i) = \prod_{j=1, j\neq i}^{q_s}(z + e_j)$ and $f_i(z) = \sum_{j=0}^{q_s-1}(b_j z^j)$. Then, $\mathcal{B}$ computes:

$$r_i' = (g^{\alpha_i} \cdot s_i^{H_1(M_i) - H_1(M_i')}, y^{\alpha_i} \cdot s_i^{x[H_1(M_i) - H_1(M_i')]}), \tag{6.2}$$

$$s_i = \prod_{j=0}^{q_s-1} (A_j)^{b_j} = (g')^{1/(x+e_i)} \quad \text{where} \quad e_i = H_1(\mathsf{CID}_i, y). \tag{6.3}$$

Since the equation holds for $g^{H_1(M_i)}(h_i \cdot y)^{\alpha_i} = g^{H_1(M_i')}(h_i \cdot y)^{\alpha_i'}$ where $h_i = g^{e_i} = g^{H_1(\mathsf{CID}_i, y)}$, and r' is the correct randomness to satisfy collision under customized identity CID_i, and public key $(g', \cdots)$. Algorithm $\mathcal{B}$ replies adversary $\mathcal{A}$ with a list of q_s new randomness $(r_1', \cdots, r_{q_s}')$.

Output: Adversary $\mathcal{A}$ wins the game by returning $(\mathsf{CID}^*, M^*, r^*, M^{**}, r^{**}, \hbar^*)$ where $(M^*, \hbar^*, r^*)$ and $(M^{**}, \hbar^*, r^{**})$ are a collision and M^{**} has never been queried during the query stage such that $g^{H_1(M^*)}(h^* \cdot y)^{\alpha_i^*} = g^{H_1(M^{**})}(h^* \cdot y)^{\alpha_i^{**}}$ where $h^* = g^{e^*} = g^{H_1(\mathsf{CID}^*, y)}$. We have:

$$r^{**} = (g^{\alpha^{**}}, y^{\alpha^{**}}) = (g^{\alpha^*} \cdot s^{*H_1(M^*)-H_1(M^{**})}, y^{\alpha^*} \cdot s^{*x\{H_1(M^*)-H_1(M^{**})\}}), \tag{6.4}$$

$$s^* = (\frac{g^{\alpha^{**}}}{g^{\alpha^*}})^{\frac{1}{H_1(M^*-H_1(M^{**}))}} = (g')^{1/(x+e^*)} = g^{f(x)/(x+e^*)}, \tag{6.5}$$

where $hk = x$ denotes the trapdoor key. We can parse f as $f(z) = \gamma(z)(z+e^*)+\gamma_{-1}$ for some $\gamma(y) = \sum_{i=0}^{q_s-1} \gamma_i z^i$ and $\gamma_{-1} \in Z_p$. Then, we can deduce by:

$$f(z)/(z+e^*) = \frac{\gamma_{-1}}{z+e^*} + \sum_{i=0}^{q_s-1} \gamma_i z^i. \tag{6.6}$$

Since $\gamma_{-1} \neq 0$ and CID^* has never been queried before (i.e. $\mathsf{CID}^* \notin \{\mathsf{CID}_1, \cdots, \mathsf{CID}_{q_s}\}$), $(z+e^*)$ cannot divide $f(z)$. So, algorithm $\mathcal{B}$ calculates:

$$\pi = \left(s^* \cdot \sum_{i=1}^{q_s}(A_i)^{-\gamma_i}\right)^{\frac{1}{\gamma_{-1}}} = g^{\frac{1}{x+e^*}}, \tag{6.7}$$

and outputs (e^*, π) where $e^* = H_1(\mathsf{CID}^*, y)$ as an answer to the q-SDHP instance $(g, g^x, \cdots, g^{x^q})$.

To bound the advantage of algorithm $\mathcal{B}$ which solves q-SDHP by using $\mathcal{A}$, refer to ref. [32]; details are omitted.

6.4.4   Efficiency Analysis of HCCH

We compare the complexity of our HCCH to the relevant schemes in Table 6.2. We set the threshold number k to 1 for a fair comparison. As it is shown in Table 6.2, our HCCH is 67% and 58% less efficient than the work in [148]. Our HCCH is acceptably efficient to be implemented in practice.

TABLE 6.2: Complexity of HCCH and other schemes

Scheme	Hash	Forge
HCCH	$5T_e + 2T_m \approx 107T_m$	$2T_e + 4T_m + 2T_i \approx 69.6T_m$
BRCB [150]	$3T_e + T_m \approx 64T_m$	$(k+1)T_e + 3T_m + T_i \geq 56.6T_m$
ICHA [219]	$4T_e + 2T_m \approx 86T_m$	$2T_e + 2T_m + T_i \approx 57.6T_m$
CHKR [149]	$3T_e + 2T_m \approx 65T_m$	$2T_e + 2T_m + T_i \approx 57.6T_m$
KEFC [148]	$3T_e + 1T_m \approx 64T_m$	$2T_e + 2T_m \approx 44T_m$

We set $k = 1$ to derive the a lower bound for the forging algorithm of [150] where k denotes the number of threshold parties to forge a collision.

6.5   Case Analysis: TUCH

In this section, we review the construction and security of our original work: Time-Updatable Chameleon Hash (TUCH) [132].

6.5.1   Construction of TUCH

We give a sketch of the construction for TUCH in Figure 6.3. We also give a brief introduction for each algorithm in TUCH scheme.

Time-Updatable Chameleon Hash (TUCH) Scheme		
System parameter: $\mathrm{param_{TUCH}} = \{G, q, g, H_1, H_2, \Delta t\}$; Public key: $hk = y = g^x$; Private key: $x \xleftarrow{R} Z_q^*$		
Hash	**Verify**	**Forge**
1. Given $\mathsf{CID} \in \{0,1\}^*$, $m \in \{0,1\}^*$, $t \in Z_q$, pick $\alpha \xleftarrow{R} Z_q^*$.	4. Given $(m, \hbar, r)$, check whether $< g, g^{\alpha t}, y, y^{\alpha t} >$ is a DH tuple.	8. Given $(m, \hbar, r)$ and x, check validity of $\hbar$.
2. Compute $h = H_1(\mathsf{CID})$ and $r = (g^{\alpha t}, y^{\alpha t})$.	5. If yes, proceed; otherwise, output $\perp$.	9. If yes, proceed; otherwise, output $\perp$.
3. Compute $\hbar = g^{\alpha t} h^{H_2(m)t}$.	6. Compute $h = H_1(\mathsf{CID})$ and check whether $\hbar \stackrel{?}{=} g^{\alpha t} h^{H_2(m)t}$.	10. Compute $r' = (g^{\alpha t} h^{((H_2(m) - H_2(m'))t}$,
	7. If yes, output 1; otherwise, output 0.	$y^{\alpha t} h^{x((H_2(m) - H_2(m'))t})$.
		11. Output r' if $< g, g^{\alpha' t}, y, y^{\alpha' t}$ is a DH tuple; otherwise, output $\perp$.

FIGURE 6.3: Sketch of TUCH Scheme

TUCH.Setup$(\lambda) \rightarrow (\mathrm{param_{TUCH}})$
TUCH.KeyGen$(\mathrm{param_{TUCH}}) \rightarrow (tk, hk)$

TUCH.Hash$(hk, \mathsf{CID}, m, t, \alpha) \to (\hbar, r)$. Parse chameleon randomness as $r = (g^{\alpha t}, y^{\alpha t})$ and chameleon hash as $\hbar = g^{\alpha t} h^{H_2(m)t}$.
   TUCH.Verify$(hk, \mathsf{CID}, (m, \hbar, r), t) \to (\bot,\ 0 \text{ or } 1)$
   TUCH.Forge$(tk, \mathsf{CID}, m', (m, \hbar, r), t) \to (r' \text{ or } \bot)$
   TUCH.Update$(tk, \mathsf{CID}, (m, \hbar, r, t), \Delta t) \to (r'' \text{ or } \bot)$

6.5.2 Security Requirement of TUCH

We capture the security of TUCH by following two security models: Experiment: $Ind_{\mathcal{A}}^{\lambda}$ and $CollRes_{\mathcal{A}}^{\lambda}$. Based on the previous studies, it shall suffice the reader to follow the security analysis.

Experiment: $Ind_{\mathcal{A}}^{\lambda}$
   $param_{\mathsf{TUCH}}^{ch} \leftarrow \mathsf{TUCH.Setup}(\lambda)$
   $(hk_{ch}, tk_{ch}) \leftarrow \mathsf{TUCH.KeyGen}(param_{\mathsf{TUCH}}^{ch})$
   $a \xleftarrow{\$} \{0, 1\}$ where $\$$ indicates random selection
   $b \xleftarrow{\$} \mathcal{A}^{\mathsf{Hash\&Forge}(tk_{ch}, \cdots, a), \mathsf{Forge}(tk_{ch}, \cdots)}(hk_{ch})$
      Here, oracle Hash&Forge on input
      $(tk_{ch}, m, m', \mathsf{CID}, \bar{t}, a)$ where $\bar{t}$ is the current time
         Set $(\hbar, r) \leftarrow \mathsf{TUCH.Hash}(hk_{ch}, \mathsf{CID}, m, \bar{t}, \alpha)$
         Set $(\hbar', r') \leftarrow \mathsf{TUCH.Hash}(hk_{ch}, \mathsf{CID}, m', \bar{t}, \alpha')$
         Set $(r'') \leftarrow \mathsf{TUCH.Forge}(tk_{ch}, \mathsf{CID}, m, (m', \hbar', r'), \bar{t})$
         If $\mathsf{TUCH.Verify}(hk_{ch}, \mathsf{CID}, m', \hbar', r', \bar{t}) = \bot \lor r'' = \bot$
            return $\bot$
         If $a = 0$:
            return $(\hbar, r)$
         If $a = 1$:
            return $(\hbar', r'')$
Return 1, if $b = a$;
else, return 0

Experiment: $CollRes_{\mathcal{A}}^{\lambda}$
   $param_{\mathsf{TUCH}}^{ch} \leftarrow \mathsf{TUCH.Setup}(\lambda)$
   $(hk_{ch}, tk_{ch}) \leftarrow \mathsf{TUCH.KeyGen}(param_{\mathsf{TUCH}}^{ch})$
   $(\mathsf{CID}^*, m^*, r^*, m^{**}, r^{**}, \hbar^*, t^*) \leftarrow \mathcal{A}^{\mathsf{Forge}'(tk_{ch}, \cdots), \mathsf{Update}(tk_{ch}, \cdots)}(hk_{ch})$
         Here, oracle Forge$'$ on input $(tk_{ch}, m, m', \mathsf{CID}, t)$:
            Return $\bot$ if $\mathsf{TUCH.Verify}(hk_{ch}, \mathsf{CID}, m, \hbar, r, t)$
            $\neq 1$
            $r' \leftarrow \mathsf{TUCH.Forge}(tk_{ch}, \mathsf{CID}, m', (m, \hbar, r), t)$
            If $r' = \bot$, return $\bot$; otherwise, return r'
         Oracle Update on input $(tk_{ch}, \mathsf{CID}, (m, \hbar, r), t), \Delta t)$:
            Return $\bot$ if $\mathsf{TUCH.Verify}(hk_{ch}, \mathsf{CID}, m, \hbar, r, t)$
            $\neq 1$
            $r'' \leftarrow \mathsf{TUCH.Update}(tk_{ch}, \mathsf{CID}, (m, \hbar, r), \Delta t)$

Return 1 if $\mathsf{TUCH.Verify}(hk_{ch}, CID^*, m^*, \hbar^*, r^*, \bar{t}) =$
$\mathsf{TUCH.Verify}(hk_{ch}, CID^*, m^{**}, \hbar^*, r^{**}, \bar{t}) = 1 \ \wedge \ m^* \neq m^{**}$
else, return 0.

6.5.3 Security Analysis of TUCH

Proof Overview 6.3 Security of TUCH mainly consists of indistinguishability and collision resistance. The informal analysis for distributing hashes proves the former. The latter is proved by the brief analysis of reduction. The reader shall find these proofs easy to prove. We leave the formal proof and bound probabilities to solve the intractability as an exercise homework for the reader.

6.5.3.1 Proof of Indistinguishability for TUCH

Indistinguishability: For sufficiently large security parameter λ, any $\mathsf{param}_{\mathsf{TUCH}} \leftarrow \mathsf{TUCH.Setup}$, $(tk, hk) \leftarrow \mathsf{TUCH.KeyGen}(\mathsf{param}_{\mathsf{TUCH}})$, all customized identities $CID \in \{0,1\}^*$, and $m, m' \in \{0,1\}^*$, denote $\mathcal{D}_{TUCH.Forge}$ and $\mathcal{D}_{TUCH.Hash}$ as distribution domains of tuple $(\hbar, r)$ where $\hbar$ and r are generated by TUCH.Hash and TUCH.Forge, respectively. Denote t as the current time and $\mathcal{R}$ as the output domains of chameleon randomness $r = (g^\alpha, y^\alpha)$. We compare the probability distributions of $\mathcal{D}_{TUCH.Hash}$ and $\mathcal{D}_{TUCH.Forge}$ as follows:

$$\mathcal{D}_{TUCH.Hash} =$$

$$\{(r, h) \mid r \in \mathcal{R}, \ m \in \{0,1\}^*, \hbar \leftarrow \mathsf{TUCH.Hash}(\mathsf{param}_{\mathsf{TUCH}}, CID, m, t, \alpha)\} \quad (6.8)$$

$$\mathcal{D}_{\mathsf{TUCH.Forge}} = \{(r', h) \mid r' \in \mathcal{R}, \ m, m' \in \{0,1\}^*, \quad (6.9)$$

$$\hbar \leftarrow \mathsf{TUCH.Hash}(\mathsf{param}_{\mathsf{TUCH}}, CID, m, t, \alpha), \quad (6.10)$$

$$r' \leftarrow \mathsf{TUCH.Forge}(tk, CID, m', (m, \hbar, t), t)\}. \quad (6.11)$$

For each chameleon hash $\hbar$, a customized identity CID and message m, there is always one-to-one correspondence $r = (\hbar \cdot h^{-H_2(m)}, \hbar \cdot h^{-H_2(m)^x})$ where $h = H_1(CID)$ and $\hbar = g^{\alpha t} h^{(H_2(m))t}$. The same applies to $(\hbar, r')$. Therefore, the probability distributions of $\mathcal{D}_{TUCH.Hash}$ and $\mathcal{D}_{TUCH.Forge}$ are indistinguishable.

6.5.3.2   Proof of Collision-resistance

Suppose $\mathcal{A}$ is a PPT adversary against the collision-resistance of our TUCH scheme. We briefly show how to construct an algorithm $\mathcal{B}$ that uses $\mathcal{A}$ to solve CDHP with non-negligible probability.

On given a CDHP instance (g, g^a, g^b), $\mathcal{B}$ interacts with $\mathcal{A}$ to output g^{ab} as follows:

At setup stage, $\mathcal{B}$ randomly chooses $x \in Z_q^*$ as the trapdoor key and computes $hk = y = g^x$ as the hash key. Set $h = g^b$ for an unknown $b \leftarrow Z_p$. $\mathcal{B}$ runs TUCH.Setup to generate $\text{param}_{\text{TUCH}}$. $\mathcal{B}$ sends $\text{param}_{\text{TUCH}}$ and hk to $\mathcal{A}$, and controls random oracles' queries $\mathcal{O}_{H_1}, \mathcal{O}_{H_2}$, Forge, and Update.

During the query stage, for each distinct query, $\mathcal{B}$ responds differently and records query histories in a list. Thus, for an identical query, $\mathcal{B}$ always return the same answer. During the response stage, for an output which satisfies a collision $(CID^*, m, r, (m', r'))$ where $r = (g^a, y^a)$ and $r = (g^{a'}, y^{a'})$, $\mathcal{B}$ computes $(y^{a'}/y^a)^{[H_2(m) - H_2(m')]^{-1}}$ as a solution to the CDHP instance. Since, $(y^{a'}/y^a)^{[H_2(m) - H_1(m')]^{-1}} = h^x$ satisfies CDHP instance (g, g^a, h, h^x), we can reduce the collision-resistance of our TUCH to the CDHP.

6.5.4   Efficiency Analysis of TUCH

We evaluate our proposed TUCH scheme's complexity by comparing to the similar schemes in Table 6.3. As it is shown in Table 6.3, our TUCH scheme is as efficient as other works in hashing, verification and forging.

TABLE 6.3: Complexity of TUCH with other schemes

Work	Algorithms			
	Hash	Verification	Forge	Update
TUCH	$3T_e + 3T_m$	$3T_e + 3T_m$	$2T_e + 4T_m$	$2T_e + 4T_m$
ACH [221]	$1T_e + 2T_m + 1T_p$	$1T_e + 2T_m + 1T_p$	$2T_e + 2T_p$	n/a
CHKE [219]	$4T_e + 2T_m$	$4T_e + 2T_m$	$2T_e + 2T_m + 1T_i$	n/a
IDCH [148]	$3T_e + 1T_m$	$3T_e + 1T_m$	$2T_e + 2T_m$	n/a

Denote T_m as group multiplication; T_e as group exponentiation; T_i as group inversion; T_p as bilinear pairing operation; n/a as not applicable.

6.6 Case Analysis: RCH

In this section, we review the construction and security of our original work: Revocable Chameleon Hash (RCH) [224].

6.6.1 Construction of RCH

We sketch our RCH in Figure 6.4, and then give a simplified algorithm for RCH.

Revocable Chameleon Hash (RCH) Scheme		
System parameter: $\text{param}_{\text{RCH}} = \{G, G_T, g, p, \hat{e}, H_1, H_2, \Delta t\}$; Public key: $y = g^x$; Private key: $x \xleftarrow{R} Z_p^*$		
Hash	**Verify**	**Forge**
1. Given CID, t, $m \in \{0,1\}^*$, compute $h = H_1(\text{CID}\|t)$.	5. Given $(\text{CID}, \hbar, r, m)$, compute $h_{t_c} = H_1(\text{CID}\|t_c)$.	9. Given $\{(\text{CID}, \hbar, r, m), x, m'\}$, compute $h = H_1(\text{CID}\|t)$ and $etd = (etd_1, etd_2) = (h^x, h^{x^2})$.
2. Pick a random $\alpha \xleftarrow{R} Z_p^*$.	6. Check whether $\hat{e}(g^\alpha, y) = \hat{e}(g, y^\alpha)$.	10. If yes, proceed; otherwise, output $\perp$.
3. Compute $r = (g^\alpha, y^\alpha)$.	7. If yes, proceed; otherwise, output 0.	11. Compute $r' = (g^{\alpha'}, y^{\alpha'}) = (g^{b_i} (etd_1)^{(H_2(m)-H_2(m'))},$ $y^\alpha (etd_2)^{(H_2(m)-H_2(m'))})$.
4. Compute $\hbar = \hat{e}(g^{\alpha t}, g) \cdot \hat{e}(h^{H_2(m)t}, y)$	8. Check whether $\hbar = \hat{e}(g^{\alpha t_c}, g) \cdot \hat{e}(h_{t_c}^{H_2(m)t_c}, y)$. If yes, output 1; otherwise, output 0.	12. Check whether: $\hat{e}(y^{\alpha'}, g) \overset{?}{=} \hat{e}(g^{\alpha'}, y)$ and $\hat{e}(g^{\alpha t}, g) \cdot \hat{e}(h^{H_2(m)t}, y) \overset{?}{=} \hat{e}(g^{\alpha' t}, g) \cdot \hat{e}(h^{H_2(m')t}, y)$. 13. If yes, output r'; otherwise, output $\perp$.

FIGURE 6.4: Sketch of RCH Scheme

RCH.Setup $(\lambda) \to (\text{param}_{\text{RCH}})$
RCH.KeyGen$(\text{param}_{\text{RCH}}) \to (x, y)$
RCH.Hash$(\text{param}_{\text{RCH}}, \text{CID}, t, y, m) \to (\hbar, r)$.
RCH.Verify$(\text{param}_{\text{RCH}}, (\text{CID}, \hbar, r, m), t_c) \to (0 \text{ or } 1)$
RCH.Td-Gen$(\text{param}_{\text{RCH}}, x, \text{CID}, t) \to (etd)$
RCH.Td-Verify$(\text{param}_{\text{RCH}}, etd, \text{CID}, t_c, y) \to (0 \text{ or } 1)$
RCH.Forge$(\text{param}_{\text{RCH}}, (\text{CID}, \hbar, r, m), etd, m', t) \to (\perp \text{ or } r')$
RCH.Un-Revoke$(\text{param}_{\text{RCH}}, etd, etd', t, k, (\text{CID}, \hbar, r, m)) \to (\perp \text{ or } r'')$

6.6.2 Security Requirement of RCH

We capture the security of our RCH by following three experiments: $Ind_{\text{RCH}}^{\mathcal{A}}(\lambda)$, $Col\text{-}NRH_{\text{RCH}}^{\mathcal{A}}(\lambda)$ and $Col\text{-}R\&NRH_{\text{RCH}}^{\mathcal{A}}(\lambda)$. In comparison with former cases, these experiments are slightly adapted according to the design of our RCH.

Experiment: $Ind_{RCH}^{\mathcal{A}}(\lambda)$

> $param_{TUCH} \leftarrow RCH.Setup(\lambda)$
> $(x_{ch}, y_{ch}) \leftarrow RCH.KeyGen(param_{TUCH})$
> Select a time $t \in_R Z_p$
> $b \xleftarrow{\$} \{0, 1\}$
> $a \leftarrow \mathcal{A}^{\mathcal{O}^{Hash\&Forge}, \mathcal{O}_{RCH}^{etd}}(param_{TUCH}, y_{ch}, t)$
> Here, oracle $\mathcal{O}_{RCH}^{etd}$ is as previously defined;
> oracle $\mathcal{O}^{Hash\&Forge}$ on input $(param_{TUCH}, x_{ch}, m, m', CID, CID', t)$
> > Set $(\hbar, r) \leftarrow RCH.Hash(\cdot, CID, m, t)$
> > Set $(\hbar', r') \leftarrow RCH.Hash(\cdot, CID', m', t)$
> > Set $(etd) \leftarrow RCH.Td\text{-}Gen(\cdot, x_{ch}, CID, t)$
> > Set $(etd') \leftarrow RCH.Td\text{-}Gen(\cdot, x_{ch}, CID', t)$
> > Set $r'' \leftarrow RCH.Forge(\cdot, etd', (CID', \hbar', m', r'), m, t)$
> > If $r'' = \bot \vee r' = \bot$, return $\bot$
> > If $b = 0$:
> > > return $(\hbar, r, CID)$
> > If $b = 1$:
> > > return $(\hbar', r'', CID')$

return 1, if $a = b$
Otherwise, return 0.

Experiment: $Col\text{-}NRH_{RCH}^{\mathcal{A}}(\lambda)$

> $param_{TUCH} \leftarrow Setup(\lambda)$
> $(x_{ch}, y_{ch}) \leftarrow RCH.KeyGen(param_{TUCH})$
> Select a time $t \in_R Z_p$
> $\leftarrow \mathcal{A}^{\mathcal{O}_{RCH}^{etd}(x_{ch}, \cdots)}(param_{TUCH}, y_{ch}, t)$
> $(CID^*, m^*, r^*, m'^*, r'^*, \hbar^*)$
> where H_1 and $\mathcal{O}_{RCH}^{etd}$ are random oracles and
> corresponding queries are recorded in $\mathcal{Q}$ and $\mathcal{L}_{Forge}$ respectively
> Output 1 if
> $1 \leftarrow RCH.Verify(param_{TUCH}, (CID^*, \hbar^*, r^*, m^*), t) \wedge$
> $1 \leftarrow RCH.Verify(param_{TUCH}, (CID^*||t, \hbar^*, r'^*, m'^*), t)$
> $\wedge CID^* \notin \mathcal{L}_{Forge} \wedge m^* \neq m'^*$

else, return 0.

Experiment: $Col\text{-}R\&NRH_{RCH}^{\mathcal{A}}(\lambda)$

> $param_{TUCH} \leftarrow Setup(\lambda)$
> $(x_{ch}, y_{ch}) \leftarrow RCH.KeyGen(param_{TUCH})$
> Denote $t_p \in_R Z_p$ as the past time
> $\leftarrow \mathcal{A}^{\mathcal{O}_{RCH}^{etd}(x_{ch}, \cdots)}(param_{TUCH}, y_{ch}, t_p)$
> $(CID^*, m^*, r^*, m'^*, r'^*, \hbar^*)$
> where H_1 and $\mathcal{O}_{RCH}^{etd}$ are random oracles and

corresponding queries are recorded in $\mathcal{Q}$
and $\mathcal{L}_{\text{Forge}}$ respectively
Output 1 if
$1 \leftarrow$ RCH.Verify($\text{param}_{\text{TUCH}}, (\text{CID}^*, \hbar^*, r^*, m^*), t_p) \wedge$
$1 \leftarrow$ RCH.Verify($\text{param}_{\text{TUCH}}, (\text{CID}^*, \hbar^*, r'^*, m'^*)$),
$\quad (t_p + k\Delta t) \wedge (\text{CID}^* \| t_c + \Delta t) \notin \mathcal{L}_{\text{Forge}} \wedge m^* \neq m'^*$
where k (> 0) is denoted as a variable of time increment;
else, return 0.

6.6.3 Security Analysis of RCH

Based on the security requirement given above, we give the security analysis.

> **Proof Overview 6.4** Security of RCH mainly consists of indistinguishability and collision resistance. The former is proved by transitions based on indistinguishability. The latter, though, is divided into two sections, and is proved by the reduction, the case analysis and deduction. Different from previous work, our RCH enables two types of hashes, including revoked one, and non-revoked one. The indistinguishability naturally applies to two kinds of hashes, however, this is not true for the collision resistance. Therefore, we give a separate analysis for each type of hash.
>
> Based on the previous analysis, readers shall not find it difficult to follow. As a hint, the proof of indistinguishability of our RCH is based on game hopping. Meanwhile, proof of collision resistance for non-revoked hash follows Boneh and Boyen's short signature scheme [32].

6.6.3.1 Proof of Indistinguishability

We briefly show how to reduce our indistinguishability to the intractability of DDHP by game hopping as follows:

Proof. Suppose $\mathcal{A}$ is an efficient adversary against our experiment $Ind^{\mathcal{A}}_{\text{RCH}}(\lambda)$. We show **Game 0** and **Game 1** as follows.

- **Game 0:** This is the original indistinguishability game as defined previously. Specifically, for $b = 1$ (where b is chosen randomly from $\{0, 1\}$ in experiment $Ind^{\mathcal{A}}_{\text{RCH}}(\lambda)$), $\hbar$ is computed as follows:

$$\hbar = \hat{e}(g^{\alpha t}, g) \cdot \hat{e}(H_1(\text{CID} \| t)^{H_2(m)t}, y). \tag{6.12}$$

- **Game 1:** It is the same as Game 0 except that it randomly samples $Z \in_R G$ and uses it to compute $\hbar$. The difference is highlighted in box as below:

$$\hbar = \hat{e}(g^{\alpha t}, g) \cdot \boxed{\hat{e}(\boxed{Z}^{H_2(m)t}, g)}. \tag{6.13}$$

Transition from Game 0 to Game 1: This hop only modifies the adversary $\mathcal{A}$'s view negligibly; otherwise, we can then construct an algorithm $\mathcal{B}$ to solve DDHP with the advantage $Adv_{\mathcal{B}}^{DDHP} = |\Pr[E_0] - \Pr[E_1]|$, where E_i is denoted as the event of $\mathcal{A}$ in winning the **Game i**. We give a proof sketch as follows:

On receiving a DDHP instance (g, g^a, g^b, g^c), $\mathcal{B}$ interacts with $\mathcal{A}$ to decide whether $c = ab \bmod p$. Concretely, $\mathcal{B}$ runs RCH.Setup to generate system parameters $\mathsf{param}_{\mathsf{TUCH}}$. It sets $y = g^b$ as the hash key where the trapdoor key $x = b$ is unknown. It samples a time t and relays $(\mathsf{param}_{\mathsf{TUCH}}, y, t)$ to $\mathcal{A}$. As previously defined, for $b = 1$, $\mathcal{B}$ generates $\hbar = \hat{e}(g^{\alpha t}, g) \cdot \hat{e}((g^c)^{H_2(m)t}, g)$. If $\mathcal{A}$ wins, $\mathcal{B}$ outputs 1; else, outputs 0.

Specifically, if $c = ab$, it is exactly the same as **Game 0**; otherwise, it is exactly the same as **Game 1**. Since Z is randomly sampled in **Game 1**, it is a one-time pad for $\mathcal{A}$ and then the adversary $\mathcal{A}$ has only a negligible advantage in winning **Game 1** (by random guess). Further, since $\mathcal{A}$ cannot distinguish between **Game 0** and **Game 1** (otherwise, it implies breaking intractability of DDHP), we can successfully reduce the indistinguishability of our RCH to the intractability of DDHP.

6.6.3.2   Collision-Resistance between Non-Revoked Hash

Proof. Suppose $\mathcal{A}$ is an efficient adversary against our $Col\text{-}NRH_{\mathsf{RCH}}^{\mathcal{A}}(\lambda)$, we briefly show how to construct an algorithm $\mathcal{B}$ to solve CDHP by interacting with $\mathcal{A}$ as below.

On given a CDHP instance (g, g^a, g^b), to derive g^{ab}, $\mathcal{B}$ proceeds as follows:

Setup. $\mathcal{B}$ runs RCH.Setup to generate $\mathsf{param}_{\mathsf{RCH}} = \{G, G_T, g, p, \hat{e}, H_1, H_2, \Delta t\}$. $\mathcal{B}$ sets $y = g^a$ as the hash key where the trapdoor key $x = a$ is unknown to $\mathcal{B}$. $\mathcal{B}$ samples an arbitrary time $t \in_R Z_p$ and sends $(\mathsf{param}_{\mathsf{RCH}}, y, t)$ to $\mathcal{A}$.

Query. $\mathcal{B}$ controls oracles H_1 and $\mathcal{O}_{\mathsf{RCH}}^{etd}$, and returns answers to $\mathcal{A}$ for each query as follows:

1 For each query on a customized identity CID_i to oracle H_1, $\mathcal{B}$ searches the tuple $((\mathsf{CID}_i \| t), Q, \epsilon, coin)$ in the history list $\mathcal{Q}$ where t denotes a previously chosen time and Q is a response, and proceeds differently as follows:

- If tuple $(\mathsf{CID} \| t, Q, \epsilon, coin)$ is found, $\mathcal{B}$ relays it to $\mathcal{A}$ directly;
- Otherwise, $\mathcal{B}$ sets $coin = 0$ or 1. We denote $Pr[E_0] = p < 1$ as probability of the event where $coin = 0$ (p will be decided later) and event $Pr[E_1] = 1 - p$ as probability for $coin = 1$. Then, $\mathcal{B}$ randomly samples $\epsilon \in_R Z_p^*$ and generates $Q = (g^b)^{coin} \cdot g^{\epsilon}$. Finally, $\mathcal{B}$ inserts tuple $(\mathsf{CID} \| t, Q, \epsilon, coin)$ in history list $\mathcal{Q}$ and returns Q to $\mathcal{A}$.

2 For each query CID_i to oracle $\mathcal{O}_{\mathsf{RCH}}^{etd}$, $\mathcal{B}$ proceeds differently as follows:

- Search the list $\mathcal{Q}$ for $\mathsf{CID}_i \| t$ (if does not exist, make a query to oracle H_1 for CID_i). Then, if $coin = 0$, $\mathcal{B}$ generates and returns $etd = (g^a)^\epsilon$ to $\mathcal{A}$.

- If $coin = 0$, aborts and terminates.

Output. For an output $(\mathsf{CID}^*, m^*, r^*, m'^*, r'^*, \hbar^*)$ returned by $\mathcal{A}$, let $(\mathsf{CID}^*, Q^*, \epsilon^*, coin^*)$ be the tuple recorded in $\mathcal{Q}$. If $coin^* = 0$, aborts and terminates; else, $\mathcal{B}$ computes an answer to the given CDHP instance by:

$$\frac{\frac{g^{b_i *}}{g^{b_i'*}}^{1/[H_2(m'^*) - H_2(m^*)]}}{(g^a)^{\epsilon^*}} = g^{ab}. \tag{6.14}$$

Specifically, if $(\mathsf{CID}^*, m^*, r^*, m'^*, r'^*, \hbar^*)$ is a valid forgery from $\mathcal{A}$, we can deduce as follows, where $h^* = H_1(\mathsf{CID}^* \| t)$:

$$\hat{e}(g^{\alpha^* t}, g) \cdot \hat{e}(h^{*H_2(m^*)t}, y) = \hat{e}(g^{\alpha'^* t}, g) \cdot \hat{e}(h^{*H_2(m'^*)t}, y). \tag{6.15}$$

Further, we have:

$$\frac{g^{b_i *}}{g^{b_i'*}}^{1/[H_2(m'^*) - H_2(m)]} = H_1(\mathsf{CID}^* \| t)^x. \tag{6.16}$$

Since $coin^* = 1$, we have $H_1(\mathsf{CID}^* \| t) = (g^b) \cdot g^{\epsilon^*}$. Immediately, we have:

$$\frac{g^{b_i *}}{g^{b_i'*}}^{1/[H_2(m'^*) - H_2(m)]} = (g^b \cdot g^{\epsilon^*})^a. \tag{6.17}$$

Thus, $\mathcal{B}$ can extract g^{ab} as below:

$$\frac{\frac{g^{b_i *}}{g^{b_i'*}}^{1/H_2(m'^*) - H_2(m)}}{(g^a)^{\epsilon^*}} = g^{ab}. \tag{6.18}$$

To bound $\mathcal{B}$'s advantage in solving CDHP, we can calculate the probability of $\mathcal{B}$'s abortion in experiment $Col\text{-}NRH_{\mathsf{RCH}}^{\mathcal{A}}(\lambda)$ by $(1-p)p^q$ where q is denoted as the number of queries made to H_1. Denote e as the base of natural logarithm, we can deduce the advantage by $\frac{1}{e(q+1)}$ if we set $p = 1 - \frac{1}{q+1}$. To explain, $\mathcal{B}$'s advantage $(1-p)p^q$ is maximized at $\frac{1}{e(q+1)}$ when $p = 1 - \frac{1}{q+1}$.

6.6.3.3 Collision-Resistance between Revoked and Non-Revoked Hash

Suppose $\mathcal{A}$ is an efficient adversary against our $\mathsf{Col\text{-}R\&NRH}^{\mathcal{A}}_{\mathsf{RCH}}(\lambda)$, we briefly show how to construct an algorithm $\mathcal{B}$ to solve CDHP by using the adversary $\mathcal{A}$. We only give a proof sketch, and the rest can be deduced from our previous analysis in section 6.6.3.2.

On a given CDHP instance (g, g^a, g^b), $\mathcal{B}$ takes g^a as the hash key where the trapdoor key a is unknown to $\mathcal{B}$. $\mathcal{B}$ relays system parameters to $\mathcal{A}$ and interacts with $\mathcal{A}$ as previously defined (given in Section 6.6.3.2). At last, $\mathcal{A}$ wins the experiment $\mathsf{Col\text{-}R\&NRH}^{\mathcal{A}}_{\mathsf{RCH}}(\lambda)$ by outputting $(\mathsf{CID}^*, m^*, r^*, m'^*, r'^*, \hbar^*)$. Then, $\mathcal{B}$ can compute

$$\left\{ [g^{b_i}{}'^* \cdot (etd_1)^{H_2(m^*)}] \cdot (g^{b_i}{}'^*)^{-1} \right\}^{H_2(m)^{-1}} = g^{ab} \tag{6.19}$$

as an answer to the CDHP instance.

To explain, we first denote t_p as the past time and, set time interval by minimum: $k = 1$. So, the current time is parsed by $t_c = t_p + \Delta t$. Based on the successful forgery made by $\mathcal{A}$ to win the experiment $\mathsf{Col\text{-}R\&NRH}^{\mathcal{A}}_{\mathsf{RCH}}(\lambda)$, we have following:

$$\hat{e}(g^{\alpha^* t_p}, g) \cdot \hat{e}(h^{* H_2(m^*) t_p}, y) =$$

$$\hat{e}(g^{\alpha'^* t_p + \Delta t}, g) \cdot \hat{e}(h'^{* H_2(m'^*) t_p + \Delta t}, y). \tag{6.20}$$

By dividing the above, we can further deduce: $H_1(\mathsf{CID}^* \| (t_p + \Delta t))^x = (etd_1'^*)^{H_2(m'^*)} = \left\{ [g^{b_i}{}'^* \cdot (etd_1)^{H_2(m^*)}] \cdot (g^{b_i}{}'^*)^{-1} \right\}^{H_2(m)^{-1}}$. Parse etd_1^* by $etd_1'^* = (g^{\alpha'^*} g^{\epsilon^*})^a$, then $\mathcal{B}$ can compute

$$\left\{ [g^{b_i}{}'^* \cdot (etd_1)^{H_2(m^*)}] \cdot (g^{b_i}{}'^*)^{-1} \right\}^{H_2(m)^{-1}} = g^{ab}. \tag{6.21}$$

as the answer to CDHP instance. We omit details due to the space limitations.

6.6.4 Efficiency Analysis of RCH

We evaluate our RCH's complexity by comparing it to other works in Table 6.4. As observed in Table 6.4, our proposed RCH is slightly inefficient in hashing, verification and forging algorithm. The main reason is for the involvement of the verification for the double-checking. Generally, our RCH is acceptably efficient to be implemented in practice.

TABLE 6.4: Complexity of RCH with other algorithms

Schemes	Algorithms					
	RCH.Hash	RCH.Verify	RCH.Forge	RCH. Etd-Gen	RCH. Td-Verify	RCH. Un-Revoke
Our RCH	$4T_e+2T_m+2T_p$	$2T_e+3T_m+4T_p$	$6T_e+8T_m+6T_p$	$2T_e$	$1T_e+1T_m+4T_p$	$12T_e+6T_m+4T_p+2T_i$
ACH [221]	$1T_e+2T_m+1T_p$	$1T_e+2T_m+1T_p$	$2T_e+2T_p$	$1T_e$	n/a	n/a
CHKE [219]	$4T_e+2T_m$	$4T_e+2T_m$	$2T_e+2T_m+1T_i$	n/a	n/a	n/a
KFCH [149]	$3T_e+2T_m$	$3T_e+2T_m$	$2T_e+2T_m+1T_i$	n/a	n/a	n/a
IDCH [148]	$3T_e+1T_m$	$3T_e+1T_m$	$2T_e+2T_m$	n/a	n/a	n/a

Denote T_m as group multiplication; T_e as group exponentiation; T_i as group inversion; T_p as bilinear pairing operation; n/a as not applicable.

6.7 Chapter Summary

This chapter generalizes and formalizes the notion of public-key hash (PKH) scheme by rigorous algorithmic design and security models of indistinguishability and collision-resistance. This chapter dives deeply into the design and analysis of current PKH schemes. Concretely, this chapter first reviews the background and information of PKH. Then, it instantiates by various PKH schemes: ACH, HCCH, TUCH and RCH which are mostly from our published works. By rich analysis of algorithmic design and proof techniques of security reduction, this chapter gives more empirical knowledge, insightful instruction and practicing sources for readers to follow. After reading this chapter, the reader should be able to: (1). Know how to define public key hash scheme and its security model. (2). Grasp the proving techniques by learning and practicing instantiated examples. (3). Give formal security analysis for any instantiated examples of PKH schemes in this chapter. To summarize, PHK is considered as the basic primitive for many other cryptographic schemes. It is an easy starting point for a beginner to follow, but can also be extended to sophisticated design. The simplicity of the instantiated cases helps the reader to quickly grasp and practice the use of proof techniques mentioned in this chapter. Thus, readers are founded with solid basics to begin their studies in the next chapter.

Chapter 7

Zero-Knowledge Proof for Blockchain

7.1 Chapter Introduction

This chapter dives deeply into the study of zero-knowledge proof system specifically applicable to blockchain. It first reviews the background and knowledge of zero-knowledge. Specifically, it shows how to build zk-SNARKs from sequential manner. This chapter then reviews three notable works of zero-knowledge proof: GS08 [225], GR16 [226] and CA07 [227] schemes. Concretely, GS08 scheme exploits bilinear maps to build efficient non-interactive zero-knowledge proof. GR16 scheme gives a tentative answer to how efficient pairing-based succinct non-interactive arguments of knowledge can be. CA17 scheme investigates two main shortcomings and countermeasures of proposals for ZKCP (a zero-knowledge proof based service). The review of the above schemes helps readers gain empirical knowledge of designing and analyzing ZKP schemes. Finally, this chapter reviews the practical implementation of zk-SNARK in Zcash, its performance and other associated problems.

After reading this chapter, you will be able to:

- Understand the meaning and definition of ZKP.

- Know generic method to construct zk-SNARKs.

- Learn basic principle to design and analyze ZKP schemes.

7.2 Introduction to ZKP

A fundamental problem in cryptography is a two-party interactive game in which a prover proves to the verifier that a predicate of a statement holdswithout letting the latter learns how to conduct the proof as the former does. A zero-knowledge protocol (ZKP) is an interactive procedure running between two principals called a prover and a verifier with the latter having a polynomially-bounded computational power. The protocol allows the former to prove that the former knows a YES answer to an NP-problem by an auxiliary input. Hence the verifier gets "zero-knowledge" about the prover's auxiliary input.

7.3   ZKP in Blockchain

In this part, we give successful use cases of ZKP and variants in blockchains, they are: NIZK in ZCoin [10] [228] and zk-SNARKs in ZCash [46].

7.3.1   Use Case 1: NIZK in ZCoin

Zerocoin [10] is an early extension of Bitcoin which exploits zero-knowledge proof. It converts Bitcoin to Zerocoin to spend such coin by proving he owns a Bitcoin. By conversion from knowledge proofs to non-interactive proofs based on Fiat-Shamir heuristic [229], it yields zero information leakage in an anonymous payment. However, is suffers from large proof sizes and verification times [230]. Garman et al. [231] extended Zerocoin [10] by decreasing the proof size and erasing the random oracle assumption. Other extensions of Zerocoin like ref. [232] and Pinocchio coin [228], require a centralized and one-time setup to initiate the parameters.

7.3.2   Use Case 2: zk-SNARK in ZCash

ZK-SNARK is a practical variant of ZKP due to its short proof size of fast verification. It has been used in Zcash [46] project to generate anonymous transactions to spend the cryptocurrencies called ZCash. Figure 7.1 illustrates the six steps that are involved in any Zcash transaction. Specifically, zk-SNARK proofs are usually generated in the Pouring Coins Phase (as highlighted in slash area in Figure 7.1).

FIGURE 7.1: Steps to Process a Zcash Transaction

7.3.3 Other Use Cases

As we earlier gave in Section 3.2.3.2, ZKP and its variants have been used in various blockchains, such as Zether [233], MimbleWimble [53], Coinjoin [234], and ZeroCoin [45]. There are also practical implementations of zk-SNARK in other areas, the reader can find them in the following published works: [228,235,236].

7.4 Introduction to of ZKP

This section gives basic knowledge and definitions associated with zero-knowledge proof. Based on preliminaries given in Section 2 and familiarity with other cryptographic primitives, we will dive deeper into zero-knowledge step-by-step by recalling important definitions whenever is necessary.

We show a detailed comparison of existing ZKP systems in Figure 7.1. To note, Libra [237] is the best among all existing systems in terms of practical prover time. It is the only system with linear prover time and succinct verification and proof size for log space uniform circuits. Libra's proof and verification are also competitive with other systems. It [237] requires a one-time trusted setup that depends on the input size. Other ZKP systems either requires a trusted setup for every statement or have deficiencies in proof size, verification time, etc.

TABLE 7.1: Comparison of existing ZKP systems

	libSNARK [236]	Ligero [238]	Bullet -proofs [239]	Hyrax [240]	libSTARK [241]	Aurora [242]	Libra [237]
$\mathcal{G}$	$\mathcal{O}(C)$ per-statement trusted setup		no trusted setup				$\mathcal{O}(n)$ one-time trusted setup
$\mathcal{P}$	$\mathcal{O}(C \log C)$	$\mathcal{O}(C \log C)$	$\mathcal{O}(C)$	$\mathcal{O}(C \log C)$	$\mathcal{O}(C \log^2 C)$	$\mathcal{O}(C \log C)$	$\mathcal{O}(C)$
$\mathcal{V}$	$\mathcal{O}(1)$	$\mathcal{O}(C)$	$\mathcal{O}(C)$	$\mathcal{O}(\sqrt{n} + d \log C)$	$\mathcal{O}(\log^2)$	$\mathcal{O}(C)$	$\mathcal{O}(d \log C)$
π	$\mathcal{O}(1)$	$\mathcal{O}(\sqrt{C})$	$\mathcal{O}(\log C)$	$\mathcal{O}(\sqrt{n} + d \log C)$	$\mathcal{O}(\log^2)$	$\mathcal{O}(\log^2)$	$\mathcal{O}(d \log C)$

Denote $\mathcal{G}$ as trusted setup algorithm; $\mathcal{P}$ as prover algorithm; $\mathcal{V}$ as verification algorithm; π as proof size respectively. Denote C as the size of circuit with depth d. n is the input size of C.

Source: Modified from Xie et al. [237], "Libra: Succinct Zero-Knowledge Proofs with Optimal Prover Computation", Annual International Cryptology Conference, Springer, Cham, (2019): 733-64.

7.4.1   Zero-Knowledge Proof and Argument

The notion of zero-knowledge $(\mathcal{ZK})$ was introduced and formalized by Goldwasser et al. [55]. It is a fundamental building block in cryptographic protocols. Zero-knowledge proofs enable a prover P to convince a verifier V of the truth of a statement without leaking any other information. To emphasize, the interaction does not yield anything to V that cannot already be computed from the input x itself in polynomial time. The central properties are captured in the notions of completeness, soundness and zero-knowledge.

Completeness: The prover can convince the verifier if the prover knows a witness testifying to the truth of the statement.

Soundness: A malicious prover cannot convince the verifier if the statement is false.

Zero-knowledge: A malicious verifier learns nothing except that the statement is true.

If a zero-knowledge protocol (P, V) for a language L requires P to have a polynomially bounded computing power, then (P, V) is called a **zero-knowledge argument** protocol. Usually, the requirement is needed in order to establish the soundness for the protocol. An argument is not as rigorous as the proof, and in particular, it fails to make a good sense when P is an unbounded entity.

Witness-indistinguishability is important security which is discussed in the next section. Briefly, it means that given two different witnesses for the statement being true, the proof reveals no information about which witness we used when we constructed the proof.

7.4.2   Non-Interactive Zero-Knowledge Proofs (NIZK)

Suppose an imaginary case where P and V both are mathematicians [200]. The former wants to travel around the world while discovering proofs for new mathematical theorems and may want to prove these new theorems to the latter in ZKP. In this scenario, the non-interactive proof is necessary because P may have no fixed address and will move away before any mail can reach it. P and V will appreciate non-interactive ZK proof.

Existing works on implementation has followed the above theoretical advances [228,235,236,243–246]. Most efficient implementations refine the quadratic arithmetic program approach of Gennaro et al. [247] and combine it with a compiler producing a suitable quadratic arithmetic program that is equivalent to the statement to be proven.

Here, we sketch one useful stronger definition of zero-knowledge. We follow the definition in [200] to capture the notion of non-interactive zero-knowledge (NIZK). In addition, we use a stronger notion of composable NIZK [248].

Definition 7.1 *(Non-Interactive Zero-Knowledge). Let R be an NP relation on pairs (x, y) with a corresponding language $\mathcal{L}_R = \{y| \ \exists \ x \ s.t. \ (x, y) \in$*

R}. *A NIZK proof system π for a relation R is a tuple of algorithms* (Π.Crs, Π.Prove, Π.Vrf) *defined as follows:*

(Π.Crs)(1^λ): *A randomized algorithm producing a common reference string* crs.

(Π.Prove)(crs, y, x): *A randomized algorithm outputting a proof π that* $R(x, y) = 1$.

(Π.Vrf)(crs, y, π): *A verification algorithm verifying whether proof π that* $y \in \mathcal{L}_R$ *is correct. If yes, output 1; otherwise, 0.*

Next, we define perfect correctness, perfect soundness, and (composable) zero-knowledge. These securities are necessary for a NIZK proof system Π to satisfy.

Definition 7.2 *(Perfect Correctness).* Π *is said to be perfectly correct if* $Experiment_{Perfect\ Correctness}$ *never returns 1.*

Definition 7.3 *(Perfect Soundness).* Π *is perfectly sound if* $Experiment_{Perfect\ Soundness}$ *never returns 1.*

Experiment$_{Perfect\ Correctness}$:
crs $\leftarrow$ TC.Key(1^λ)
$(y, x) \leftarrow \mathcal{A}(\text{crs})$
$\pi \leftarrow \Pi$.Prove(crs, y, x)
return 1, if $(x, y) \in R$ and
Π.Vrf(crs, y, π) $= 0$;
else, return 0.

Experiment$_{Perfect\ Soundness}$:
crs $\leftarrow$ TC.Key(1^λ)
$(y, \pi) \leftarrow \mathcal{A}(\text{crs})$
return 1, if
Π.Vrf(crs, y, π) $= 1$ and $y \notin \mathcal{L}_R$;
else, return 0.

To define zero-knowledge, we need to specify two randomized algorithms. The first one, Π.SimCrs(1^λ), takes a security parameter and generates a simulated common reference string with a corresponding trapdoor key tk. The second, a simulator, Π.Sim(cry, y, tk) takes as input a statement and the trapdoor key, but no any witnesses, and outputs a proof π for which Π.Vrf(crs, y, π) accepts. Note, the simulation algorithm can produce proofs for any statement since it does not check whether $y \in \mathcal{L}_R$ or not.

Definition 7.4 *((Composable) Zero-Knowledge).* Π *is said to be zero-knowledge if, both experiments defined in next return 1 with probabilities $\frac{1}{2}$ + $negl(\lambda)$, where negl denotes a negligible function. If we require $Exp_{SIM-IND}$ to return 1 with probability $\frac{1}{2}$, we say that Π has a perfect zero-knowledge simulation.*

Experiment$_{CRS-IND}$:
$b \leftarrow \{0, 1\}$
if $b = 0$, crs $\leftarrow \Pi$.Crs(1^λ)
else, (crs, tk) $\leftarrow \Pi$.SimCrs(1^λ)
$b^* \leftarrow \mathcal{A}(\text{crs})$

return 1 if $b = b^*$ and 0 otherwise.
Experiment$_{SIM-IND}$:
(crs, tk) $\leftarrow \Pi$.SimCrs(1^λ)
$(y, x, state) \leftarrow \mathcal{B}(\text{crs})$
$\pi_0 \leftarrow \Pi$.Prove(crs, y, x)

$\pi_1 \leftarrow \Pi.\mathsf{Sim}(\mathsf{crs}, y, \mathsf{tk})$

$b \leftarrow \{0, 1\}$

$b^* \leftarrow \mathcal{B}(state, \pi_b, \mathsf{tk})$

return 1 if $b = b^*$ and $(x, y) \in R$;

otherwise, return 0.

7.4.3 Zero-Knowledge Succint Non-Interactive Arguments of Knowledge (zk-SNARK)

zk-SNARKs are practical variants of non-interactive zero-knowledge proofs. NIZKs requires a one-time trusted setup for generating a common reference string for NIZKs. Similarly, zk-SNARKs also requires a one-time trusted setup for the proving and the verification keys. The difference lies in efficiency guarantees. In a NIZK, the proof length and the verification time depend on the NP-language being proved. Conversely, in a zk-SNARK, proof length depends only on the security parameter, and the verification time depends only on the instance size. Thus, zk-SNARKs is considered as succinct NIZKs with short proofs and fast verification times.

Formally, let $\mathcal{L}$ be an NP-language, and let C be a non-deterministic decision circuit for $\mathcal{L}$ on a given instance size n. A zk-SNARK can be used to prove and verify membership in $\mathcal{L}$, for instances of size n. For example, asking C as an input, a trusted party conducts a one-time setup phase to derive two public keys: a proving key pk and a verification key vk. The proving key pk allows any prover to produce a proof π attesting to the fact that $x \in \mathcal{L}$, for instance, x (of size n) of his choice. Anyone can use the verification key vk to check the validity of the proof π. Besides, zk-SNARK proofs are publicly verifiable: anyone can verify π, without ever having to interact with the prover who generated π. Further details can be found in [236,249] for zk-SNARK.

7.4.4 Known constructions and security of zk-SNARK

There are many zk-SNARK constructions in the literature, e.g. [228,235, 236,247,249–252]. We are interested in the following zk-SNARKs:

- Arithmetic circuit satisfiability. The most efficient ones for this language are based on quadratic arithmetic programs [228,235,236,247,249].

- Based on knowledge-of-exponent assumptions and variants of Diffie-Hellman assumptions. Related works are [159,250,253].

Particularly, we will exemplify GS08 [225], GR16 [226], and CA17 [227] schemes in this chapter as case analyses.

7.5 Steps to Achieve a zk-SNARK

Buterlin explained the method to achieve zk-SNARK in ref. [254]. Ref. [255] also gives a generic method to achieve zk-SNARK.

Briefly, zk-SNARK is designed to create a function or a protocol that takes the proof from a prover P, and enables a verifier V to check its validity. Concretely, to verify a zk-SNARK proof one proceeds as follows. The computation is first turned into an arithmetic circuit. Each of its wires is then assigned a value that results from feeding specific inputs to the circuit. Next, each computing node of the arithmetic circuit (also called a "gate") is turned into a constraint that verifies the output wire that has the value assigned to the input wires. This process involves transforming statements into the formats on which a zk-SNARK proof can be performed. We give a sketch of the following steps for achieving a zk-SNARK in Figure 7.2.

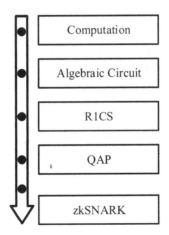

FIGURE 7.2: Steps to Achieve zk-SNARK

To apply zk-SNARKS to a problem. First, convert the problem into the right "form". This form is called a quadratic arithmetic program (QAP). Then, transform the code of a function into a highly non-trivial one. We exemplify it by asking the reader to give a solution to a cubic equation: $x^3 + x + 5 = 35$ (hint: the answer is 3).

7.5.1 Computation to Algebraic Circuit

The first step is to convert the original code into a sequence of statements that are of two forms $x = y$ and $x = y(op)z$ (where op can be $+, -, *, /$, and y, z can be variables, numbers). You can think of each of these statements as being kind of logic gates in a circuit. The result is as shown in code$_2$). Obliviously, these two codes are equivalent.

code$_1$
```
def (x) :
    y = x³
    return x + y + 5
```

code$_2$
```
sym₁ = xˣ
y = sym₁ˣ
sym₂ = y + x
out = sym₂ + 5
```

Source: Data from Buterin [254], "Quadratic Arithmetic Programs: from Zero to Hero", Medium, 2016.

7.5.2   Algebraic Circuit to R1CS

We convert the above into something called a rank-1 constraint system (R1CS). Example and instantiation of R1CS are given in ref. [254].

7.5.2.1   What's Arithmetic Circuits?

To explain, an arithmetic circuit consists of gates computing arithmetic operations, like addition and multiplication, with wires connecting the gates. We show an example of an algebraic circuit in Figure 7.3.

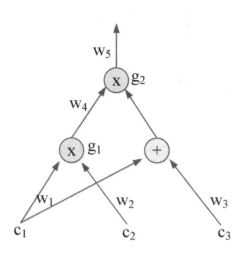

FIGURE 7.3: Arithmetic Circuits

Source: Data from Banerjee et al. [255], "Demystifying the Role of zk-SNARKs in Zcash", arXiv preprint arXiv: 2008. 00881, 2020.

A circuit is composed of wires, which performs an operation on input and returns outputs. In complexity theory, it is common to consider logical circuits, where the value of each wire is either 0 or 1, and gates perform logical operations like AND, OR, NOT, and their variants. The gates of a circuit must be organized so that computation always flows in a single direction from inputs to outputs. Computations are flattened and laid out in a way that each gate is evaluated only once and the wires never change the value. One can instantaneously determine the value of all wires. The bottom wires are the input wires, and the top wire is the output wire which returns the result of the circuit computation on the inputs.

7.5.2.2 Rank 1 Constraint System (R1CS)

R1CS is a set of constraints specified by 3 linear combinations, called a, b, c. Given an assignment describing the full state of a computation, the R1CS can be used to check that all the steps were correctly computed. The solution to an R1CS is a vector s, where s must satisfy the equation $s \cdot a * s \cdot b - s \cdot c = 0$. To explain, if we "zip together" a and s, then do the same to b and s, and then c and s, the third result equals the product of the first two results. We give an example in Figure 7.4. Further details can be found in ref. [254,255].

FIGURE 7.4: An Example of R1CS

Source: Data from Banerjee et al. [255], "Demystifying the Role of zk-SNARKs in Zcash", arXiv preprint arXiv: 2008. 00881, 2020.

7.5.3 R1CS to QAP

In this section, we show the transformation from R1CS into QAPs.

7.5.3.1 Quadratic Arithmetic Program (QAP)

In 2013, Gennaro, Gentry, Parno and Raykova [247] defined the quadratic arithmetic program (QAP) as an extremely useful translation of computations into polynomials. QAPs have become the basis for zk-SNARK constructions and have been used by Zcash.

Suppose Alice wants to prove to Bob that she knows $c_1, c_2, c_3 \in \mathbb{F}_p$ such that $(c_1 \cdot c_2) \cdot (c_1 + c_3) = 7$. Following the idea discussed in Section 7.5.1, we next show the transformation from c_1, c_2, c_3 to an arithmetic circuit. Once having the full set of constraints, we can create the QAP. Our goal is to devise a set of polynomials that simultaneously encode all of the constraints. One smart trick is to build the polynomials in a way that they can generate all of the constraints. To do so, we employ a primitive called: Lagrange interpolation [256]. Doing a Lagrange interpolation on a set of points (i.e. (x, y)) implies a polynomial that passes through all of those points. So, to convert given R1CS into QAP, one goes from four groups of three vectors of length six to six groups of three degree-3 polynomials, where evaluating the polynomials at each x-coordinate represents one of the constraints. The use of Lagrange interpolation achieves this goal. For more details, refer to ref. [254,255].

7.6   Case Analysis: GS08 Scheme

This section reviews the construction and the security of an article (GS08) by Groth and Sahai [225]. Groth and Sahai focus on popular tools called groups with bilinear maps to construct NIZK. Their primary motivation is to pursuit a large class of practical cryptographic protocols based on bilinear groups.

7.6.1   Introduction to GS08 Scheme

In this work, Groth and Sahai use XDH and SXDH assumptions (as specified by Definition 2.12 and Definition 2.13 in Section 2.5.2) to construct efficient non-interactive proof systems for bilinear groups. The core tool is non-interactive witness-indistinguishable proofs. The proof should be complete and sound. Meanwhile, the proposed non-interactive proofs should also satisfy perfect witness-indistinguishability. These are captured by security requirements and proven via rigorous analysis.

7.6.2   Definition and Security Requirement of GS08 Scheme

Before constructing the scheme, Groth and Sahai formalize several security requirements: Perfect completeness; Perfect soundness; Composable witness indistinguishability; Composable zero-knowledge. We sketch the security requirements of ref. [225] in Figure 7.5.

To note, Perfect Witness-Indistinguishability is stronger security than Composable Witness-Indistinguishability. Here, we leave the reader to finish the formalization of this security. The reader shall find complete answer in ref. [225]

Perfect Witness-Indistinguishability. This means that given two different witnesses for the statement being true, the proof reveals no information about which witness we used to construct the proof.

Security Requirement of NIWI

Perfect Completeness.

1. For all $\mathcal{A}$, $\Pr[\sigma \leftarrow K(1^k); (x,w) \leftarrow \mathcal{A}(\sigma); \pi \leftarrow P(\sigma, x, \pi) = 1$ if $(\sigma, x, w) \in R] = 1$

Perfect Soundness.

2. For all $\mathcal{A}$, $\Pr[\sigma \leftarrow K(1^k) : \mathcal{A}(\sigma) = 1] \approx \Pr[(\sigma, \tau) \leftarrow S_1(1^k) : \mathcal{A}(\sigma) = 1]$

Composable Witness Indistinguishability.

3. For all $\mathcal{A}$ $\Pr[\sigma \leftarrow K(1^k) : \mathcal{A}(\sigma) = 1] \approx \Pr[\sigma \leftarrow S(1^k) : \mathcal{A}(\sigma) = 1]$

$\Pr[\sigma \leftarrow S(1^k); (x, w_0, w_1) \leftarrow \mathcal{A}(\sigma); \pi \leftarrow P(\sigma, x, w_0) : \mathcal{A}(\pi) = 1]$

$= \Pr[\sigma \leftarrow S(1^k); (x, w_0, w_1) \leftarrow \mathcal{A}(\sigma); \pi \leftarrow P(\sigma, x, w_1) : \mathcal{A}(\pi) = 1]$

Composable Zero-Knowledge.

4. For all $\mathcal{A}$ $\Pr[\sigma \leftarrow K(1^k) : \mathcal{A}(\sigma) = 1] \approx \Pr[(\sigma, \tau) \leftarrow S_1(1^k) : \mathcal{A}(\sigma) = 1]$,

$\Pr[(\sigma, \tau) \leftarrow S_1(1^k); (x, w) \leftarrow \mathcal{A}(\sigma, \tau); \pi \leftarrow P(\sigma, x, w) : \mathcal{A}(\pi) = 1 \text{and} (x, w) \in R]$

$= \Pr[(\sigma, \tau) \leftarrow S_1(1^k); (x, w) \leftarrow \mathcal{A}(\sigma, \tau); \pi \leftarrow S_2(\sigma, \tau, w) : \mathcal{A}(\pi) = 1 \text{and} (x, w) \in R]$

FIGURE 7.5: Security Requirements of GS08 Scheme

7.6.3 Analysis of GR08 Scheme

At a high level, Groth and Sahai construct witness-indistinguishable proof via a protocol proceeded by the following steps:

1. Commit to all variables.

2. Generate a proof for each pairing product equation.

3. Generate a proof for each multi-exponentiation relationship.

4. Generate a NIWI proof for each general arithmetic gate.

Groth and Sahai arrange several sections with instantiations in their work for demonstration, they are:

1. Commitment from Modules.

2. Setup for generating CRS.

3. Pairing product equations.

4. General arithmetic gates.

5. Multi-exponentiation.

In each of the above section, Groth and Sahai consider all mathematical operations that could take place in the context of a bilinear group exponentiation, addition or multiplication of exponents, multiplication of group elements and use of the bilinear map. Briefly, given parameters for initiating relevant building blocks, and equations over variables, Groth and Sahai's protocol works to output a witness-indistinguishable proof.

1. Utilize Commitment from Modules (Sec 4) to commit to a ring element for each equation.

2. Based on committed variables, run Pairing product equation ([225], Sec 5) to construct non-interactive proofs for the committed values satisfying a pairing product equation.

3. Run to General arithmetic gates ([225], Sec 6) make a NIWI proof for the general arithmetic gate using the NIWI proof.

4. Run Multi-exponentiation ([225], Sec 7) to derive NIWI proof accordingly.

Following the above idea, we give a sketch of constructing a witness indistinguishable proof under XDH and SXDH assumptions in Figure 7.6.

7.6.4 Security Analysis of Groth and Sahai's Scheme

Proof Overview 7.1 The proving techniques used here are proof by theorem and illation. Lemma 7.1 defines the preconditions to capture the equality of distributions of the resulting proofs. To prove this lemma, Groth and Sahai conclude a uniform random sample of two proofs under a restriction by illation. This process is logical and requires the reader to be familiar with all associated parameters.

In this part, we analyze the security of this work. Since Groth and Sahai proposed many building blocks and a protocol to generate a non-interactive witness-indistinguishable proof, it is needed to prove each building block's security. Here, we focus on the security of pairing production equation. In other

Instantiation of witness indistinguishable proof under XDH and SXDH	
System parameter: $(p, G_1, G_2, G_T, \hat{e}, g_1, g_2 \leftarrow \mathcal{G}(1^k))$.	
CRS Generation.	**Simulated CRS Generation.**
1. Pick $x_1, x_2, y_1, y_2, r, s \leftarrow Z_p^*$ and $t_1, t_2 \leftarrow Z_p$.	4. Pick $x_1, x_2, y_1, y_2, r, s \leftarrow Z_p^*$ and $t_1, t_2 \leftarrow Z_p$.
2. Compute u_1, u_2, u and v_1, v_2, v.	5. Compute u_1, u_2, u and v_1, v_2, v differently from step 2.
3. Set $\sigma = (p, G_1, G_2, G_T, \hat{e}, g_1, g_2, u_1, u_2, u, v_1, v_2, v)$.	6. Set $\sigma = (p, G_1, G_2, G_T, \hat{e}, g_1, g_2, u_1, u_2, u, v_1, v_2, v)$.
Proof.	**Verify**
7. Given $x_m \in G_1, y_n \in G_n, \phi_k, \theta_l \in Z_p$, pick $r_{m_i}, s_{n_j}, \rho_k, \sigma_l \leftarrow Z_p$.	9. Check the proof with respect to each euqation
8. Commit as c_k, d_k, c'_k, d'_k.	

FIGURE 7.6: Instantiation of Constructing a Witness Indistinguishable Proof under XDH and SXDH

words, we focus on proving the security of non-interactive proofs generated from committed values satisfying a pairing product equation.

Groth and Sahai give four lemmas to state the security achieved by proof generated from each proposed scheme. We abbreviate Lemma 1 in [225] as follows:

Lemma 7.1 *Assume we have* $u_1, \cdots, u_I \in M_1$ *and* $v_1, \cdots, v_J \in M_2$ *and* $\eta_1, \cdots, \eta_H$ *generating the kernel of* μ. *Consider two witnesses* x_q, y_q, r_{qi}, s_{qj} *and* $x'_q, y'_q, r'_{qi}, s'_{qj}$ *satisfying the equations. If for all* q *we have* $x_q, x'_q \in U$, $y_q, y'_q \in V$ *and we pick the* t_{ij}'s *and* t_h's *at random from* R, *then the distribution of the resulting proofs* π_i, $\rho_j s$ *and* π'_i, ρ'_j *are identical.*

The Lemma 7.1 states the condition under which the proofs' distribution is supposed to be identical. To prove this lemma, the author just enumerated the parameters associated with the proof and discussed them by reasoning and summarization. The proof of other lemmas follows similar methodologies.

Proof of Lemma 7.1. Consider a witness x_q, y_q, r_{qi}, s_{qj} as specified in the Lemma 7.1. This gives us $\pi_1, \cdots, \pi_I \in V$ and $\psi_1, \cdots, \psi_J \in U$. Since we pick the t_{ij}'s at random, the π_i's are distributed uniformly at random in V. Consider any fixed tuple $(\pi_{i1}, \cdots, \pi_I)$ of elements from V.

The corresponding ψ_j's in U satisfy $\prod_{j=1}^J E(\psi_j, v_j) = \prod_{q=1}^Q E(c_q, d_q) \cdot T^{-1} \prod_{i=1}^I E(u_i, \pi_i)^{-1}$. Since $\eta_1, \cdots, \eta_H$ generate the kernel of μ, by picking the t_h's at random in the construction of ψ_j's, we get random ψ_j's from U such that $\prod_{j=1}^J E(\psi_j, v_j) = \prod_{q=1}^Q E(c_q, d_q) \cdot T^{-1} \prod_{i=1}^I E(u_i, \pi_i)^{-1}$.

We conclude that with the witness x_q, y_q, r_{qi}, s_{qj} we get a uniform random sample of π_i and ψ_j under the restriction that $\prod_{q=1}^{Q} E(c_q, d_q) \cdot T \prod_{i=1}^{I} E(u_i, \pi_i) \cdot \prod_{j=1}^{J} E(\psi_j, v_j)$. By a similar argument, the other witness, $x_q', y_q', r_{qi}', s_{qj}'$ gives exactly the same distribution on π_i', ψ_j'.

7.7 Case Analysis: GR16 Scheme

In this section, we review the GR16 scheme proposed by Groth in [226]. This work's motivation is seeking appropriate tools to turn the circuit into a QAP for QAP-based zk-SNARKs. As contributions, he proposed two constructions. The first contribution is a pairing-based SNARK for arithmetic circuit satisfiability. The second contribution seeks to find lower bound for pairing-based non-interactive arguments.

7.7.1 Non-Interactive Zero-Knowledge Arguments of Knowledge

To start, we first review definitions for non-interactive zero-knowledge arguments of knowledge.

Let $\mathcal{R}$ be a relation generator which on given a security parameter λ, outputs a polynomial time decidable binary relation r. For pairs $(\phi, w) \in \mathcal{R}$ we call ϕ the statement and w the witness.

We define $\mathcal{R}_\lambda$ to be the set of possible relations. An efficient publicly verifiable non-interactive argument for R is a quadruple of probabilistic polynomial algorithms (Setup, Prove, Vfy, Sim) as sketched in Figure 7.7.

Non-Interactive Zero-Knowledge Arguments of Knowledge

$(\sigma, \tau) \leftarrow$ Setup(λ): The setup produces a common reference string σ

and a simulation trapdoor τ for the relation R.

$\pi \leftarrow$ Prove(R, σ, ϕ, w): The prover algorithm takes as input a common reference string σ

and $(\phi, w) \in R$ and returns an argument π.

$0/1 \leftarrow$ Vfy(R, σ, ϕ, π): The verification algorithm takes as input a common reference string σ

a statement ϕ and an argument π and returns 0 (reject) or 1 (accept).

$\pi \leftarrow$ Sim(R, τ, ϕ): The simulator takes as input a simulation trapdoor

and statement ϕ and returns an argument π.

FIGURE 7.7: Non-Interactive Zero-Knowledge Arguments of Knowledge

7.7.2 Construction of Groth's GR16 Scheme

In the GR16 scheme [226], Groth gives a construction of pairing-based NIZK argument for QAP in two steps.

Step 1. Groth gives the construction of non-interactive linear proofs (NILP). It is based on the definition of NILP and a methodology to compile any NILPs (Setup, Prove, Vfy, Sim) to publicly verifiable non-interactive arguments (Setup', Prove', Vfy', Sim'). Concretely, the compiler methodology exploits a variant of the Paillier encryption [249]. Groth assumes a primitive called "a split NILP" where the common reference string and corresponding proof are split into two parts. Concretely, a function called ProofMatrix is summoned to generate a matrix $\Pi = (\begin{pmatrix} \Pi_1 & 0 \\ 0 & \Pi_2 \end{pmatrix})$. Such matrix Π is used to compute the proof π. In addition, a function $\mathsf{Test}(R, \phi)$ is summoned to get an arithmetic circuit and determine the validity of the proof. Later, Groth instantiated the details of compiler for transforming (Setup, Prove, Vfy, Sim) to (Setup', Prove', Vfy', Sim').

While formalizing the compiler, Groth specifies a potential attack where the prover sees the common reference string and extracts useful information to form her matrix Π. Groth formalizes the definition by a notion called disclosure-free NILP. This definition seeks to capture the security of the common reference string's testing results being leaked from independently generated witnesses.

Step 2. Based on step 1, Groth formalizes the security for non-interactive arguments from the compiler (Lemma 1, ref. [226]). Groth parses security by perfect completeness and perfect zero-knowledge, and analyzes accordingly. Specifically, Groth captures the disclosure-freeness by adversary's negligible probability in learning information about CRS. Further, Groth suggests defining modified adversary who does not make any zero tests on CRS instead of making queries during the simulation. This equals adversary who chooses matrices independently. Finally, Groth concludes by arguing the adversary's negligible probability of winning in associated game.

7.7.3 Analysis of GR16's First Construction

Following the basic idea given in Section 7.7.2, we review the concrete constructions given by Groth. A sketch of the first construction is given in Figure 7.8. Here, Groth seeks to construct a NILP for quadratic arithmetic program generators that output relations of the form.

To prove the security of the first construction, Groth defines Theorem 1 in ref. [226] to capture three security models desired. To prove Theorem 1, Groth focuses on demonstrating the extraction of a witness from an adversary attacking this security. Groth enumerates the stages and parameters associated with each algorithm in the scheme to perform strict security analysis. Specifically, Groth uses Schwartz-Zippel's lemma [257] to capture the prover's

---

Non-Interactive Linear Proofs (NILP) for Quadratic Arithmetic Programs

System parameter: $\mathsf{param}_{\mathsf{NILP}} = \{p, G_1, G_2, G_T, e, g, h\}$.

Relation: $R = (\mathbb{F}, aux, l, \{u_i(X), v_i(X), w_i(X)\}_{i=0}^{m}, t(X))$.

$(\sigma, \tau) \leftarrow \mathsf{Setup}(R)$:

1. Given R, pick $\alpha, \beta, \gamma, \delta, x \leftarrow \mathbb{F}^*$.
2. Set $\tau = (\alpha, \beta, \gamma, \delta, x)$.

3. Set σ as specified in construction.

$\pi \leftarrow \mathsf{Prove}(R, \sigma, a_1, \cdots, a_m)$:

4. Pick $r, s \leftarrow \mathbb{F}$.

5. Compute a $3 \times (m + 2n + 4)$ matrix Π.

6. Generate $\pi = \Pi$, $\sigma = (A, B, C)$.

$0/1 \leftarrow \mathsf{Vfy}(R, \sigma, a_1, \cdots, a_l)$:

7.   Compute t such that $t(\sigma, \pi) = 0$.
8.   Output 1 if the test passes (see test in construction); other-wise, 0.

$\pi \leftarrow \mathsf{Sim}(R, \tau, a_1, \cdots, a_l)$:

9. Pick $A, B \leftarrow \mathbb{F}$.

10. Compute C.

11. Output $\pi = (A, B, C)$.

FIGURE 7.8: Sketch of NILP Scheme

success probability in the verification. Based on this, Groth assumes that $B_a = 0$. This, in turn, yields a further assumption $A_\alpha = B_\beta = 1$. Next, Groth makes a further assumption on the simplified A and B and the corresponding equations. From progressive assumption and deducing, Groth derives the witness for the statement. This completes the proof.

7.7.4   Analysis of Groth's Second Construction

The second construction proposed by Groth achieves a pairing-based NIZK argument for the quadratic arithmetic programs. We give a sketch of the construction in Figure 7.9. Based on the compiler technique Groth previously proposed, we can derive non-interactive linear proofs (NILP) from this construction.

An important design feature of the NILP we gave above is that it can easily make it a split NILP. The proof elements A, B and C are only used once in the verification equation, and therefore, it is easy to assign them to different sides of the bilinear test. By splitting the common reference string into two parts that enable the computation of each side of the proof, we then

NIZK Arguments for QAP

System parameter: $\text{param}_{\text{NILP}} = \{p, G_1, G_2, G_T, e, g, h\}$.

Relation: $R = (p, G_1, G_2, G_T, \hat{e}, g, h, l, \{u_i(X), v_i(X), w_i(X)\}^m$

$(\sigma, \tau) \leftarrow \text{Setup}(R)$:	$\pi \leftarrow \text{Prove}(R, \sigma, a_1, \cdots, a_m)$:
1. Given R, pick $\alpha, \beta, \gamma, \delta, x \leftarrow Z_p^*$.	4. Pick $r, s \leftarrow Z_p$.
2. Set $\tau = (\alpha, \beta, \gamma, \delta, x)$.	5. Compute $\pi = ([A]_1, [C]_1, [B]_2)$ as specified in detailed construction.
3. Compute $\sigma = ([\sigma_1]_1, [\sigma_2]_2)$ as specified in detailed construction.	
$0/1 \leftarrow \text{Vfy}(R, \sigma, a_1, \cdots, a_l, \pi)$:	$\pi \leftarrow \text{Sim}(R, \tau, a_1, \cdots, a_l)$:
6. Parse $\pi = ([A]_1, [C]_1, [B]_2) \in G_1^2 \times G_2$.	8. Pick $A, B \leftarrow Z_p$.
7. Determine acceptance according to equation.	9. Compute π ($[A]_1, [C]_1, [B]_2$) with C.

FIGURE 7.9: Sketch of NIZK Arguments for QAP

get a split NILP. The returned split NILP is also disclosure-free and therefore can be compiled into a NIZK argument in the generic group model [226].

Similarly, Groth announces the security achieved by the proposed construction in Theorem 2 in ref. [226]. It mainly remains to prove the disclosure-freeness of CRS. Groth performs a case analysis and prediction on the Laurent polynomials contained in witness. Then, he applies the Schwartz-Zippel lemma extension to one of the cases to analyse the winning probability. The typical proof techniques Groth used in this section is case analysis.

7.8 Case Analysis: CA17 Scheme

In this section, we review the construction and countermeasures given in CA17 scheme by Campanelli et al. [227]. The main contributions of this work are showing attacks that a buyer (user) learns partial information via the revelation of two main shortcomings of current proposals. Specifically, they studied zero-knowledge contingent payment (ZKCP) protocols. It allows for a fair exchange of sold goods and payments over the Bitcoin network.

7.8.1   Prepare to Attack: Building ZK-SNARKs from QAP

In this section, Campanelli et al. [227] pointed out how to build zk-SNARKs from quadratic arithmetic programs. Before analyzing the construction, we briefly review the notion of quadratic arithmetic programs (QAPs) in Definition 7.5 and show how to use it to build zk-SNARKs by following the idea of [247]. We give a sketch of our construction in Figure 7.10.

Definition 7.5 *A QAP Q over a field $\mathbb{F}$ is defined by three sets of polynomials $A : \{A_i(x)\}_{i=0}^m$, $B : \{B_i(x)\}_{i=0}^m$, $C : \{C_i(x)\}_{i=0}^m$ and a target polynomial $Z(x)$. If we take a function $f : \mathbb{F}^n \to \mathbb{F}^{n'}$, then we say that Q computes f, if given a valid assignment $(c_1, \cdots, c_{n+n'})$ of inputs and outputs of f, there exist coefficients $(c_{n+n'+1}, \cdots, c_m)$ such that $Z(x)$ divides the following polynomial*

$$p(x) := (A_0(x) + \sum_{k=1}^m c_k \cdot A_k(x)) \cdot (B_0(x) + \sum_{k=1}^m c_k \cdot B_k(x)) +$$

$$-(C_0(x) + \sum_{k=1}^m c_k \cdot C_k(x)) \qquad (7.1)$$

Briefly, there should exist a polynomial $H(x)$ such that $p(x) = H(x) \cdot Z(x)$. We refer to m and the degree of $Z(x)$ as the size and the degree of Q, respectively.

Source: Data from Campanelli et al. [227], "Zero-Knowledge Contingent Payments Revisited: Attacks and Payments for Services", CCS 2017, Dallas Texas USA October, 2017, pp. 229–243.

To build a QAP for a function f, one shall use an arithmetic circuit $\mathcal{C}$ to represent f. Then, pick a distinct root r_g for any of its multiplicative gates. Next, build the target polynomial as $Z(x) : \prod_g (z - r_g)$, and we label each input of the circuit and each output of a multiplicative gate with an index $i \in [m]$. Define the polynomials A, B, C to respectively encode each gate's left, right and output wire. So, for any gate, g and its root r_g, the condition above can be seen as:

$$(\sum_{k=1}^m c_k \cdot A_k(r_g)) \cdot (\sum_{k=1}^m c_k \cdot B_k(r_g))$$

$$= (\sum_{k \in I_L} c_k \cdot A_k(r_g)) \cdot (\sum_{k \in I_R} c_k \cdot B_k(r_g)) = c_g C_k(r_g) = c_g \qquad (7.2)$$

Following the idea given in [236], we can now use QAPs to build zk-SNARKs.

Attacks on ZKCP with untrusted CRS

System parameter: $\text{pp} := (r, e, \mathcal{P}_1, \mathcal{P}_2, G_1, G_T)$.

Circuit $\mathcal{C}: \mathbb{F}_r^n \times \mathbb{F}_r^h \to \mathbb{F}_r^l$.

$(pk, vk) \leftarrow$ KeyGeneration$(\mathcal{C})$

1. Compute (A, B, C, Z) and extend A, B, C accordingly.

2. Pick $\tau, \varphi_A, \varphi_B, \alpha_A, \alpha_B, \alpha_C, \beta, \gamma \xleftarrow{\$} \mathbb{F}_r$.

3. For $i = 0, \cdots, m + 3$, set $\{pk_{A,i}, pk'_{A,i}, \cdots$. Set pk

4. Set $\{vk_A, vk_B, \cdots\}$, Set vk.

5. Output (pk, vk).

Verifier

$0/1 \leftarrow$ Verify(vk, x, π):

12. Compute vk_x.

13. Check QAP divisibility.

14. Out 0 or 1.

Prover

$\pi \leftarrow$ Prove$(R, \sigma, a_1, \cdots, a_m)$

6. Compute (A,B,C,Z) with $\mathcal{C}$.

7. Compute the QAP witness $s \in \mathbb{F}^m$.

8. Pick $\delta_1, \delta_2, \delta_3 \xleftarrow{\$} \mathbb{F}_r$.

9. Compute the polynomial $H(z)$ with $A(z), B(z), C(z)$.

10. Set pk_A, pk'_A
11. Compute and output $\pi :=$
$(\pi_A, \pi'_A, \pi_B, \pi'_B, \pi_C, \pi'_C, \pi_K, \pi_H)$

FIGURE 7.10: Sketch of ZK-SNARKs from QAP

7.8.2 Potential Attacks: Learning Information by modifying the CRS

As discussed in CA17 scheme [227], a variety of attacks can be launched to extract information about the Sudoku solution during the offline phase of the ZKCP if a malicious verifier is allowed to set the CRS. We briefly introduce them as follows:

Changing the Circuit: This is potentially the easiest attack to observe. A malicious verifier can replace the CRS with the QAP encoding of a modified function f for attack, for example, modifying the sets of polynomials A, B, C and C to $(\tilde{A}, \tilde{B}, \tilde{C}, \tilde{Z})$.

Learning one wire is sufficient: Learning the value of the wire w_j will allow us to determine a particular cell in the "Pay to Sudoku" (PtS) implementation [258].

Choosing τ as one of the roots of Z: If one selects τ as one of the roots of $Z(x)$, then the leakage of component τ_B reveals the value $\gamma_j = c_j \phi_B \mathcal{P}_2$.

This leakage allows recovering c_j. This attack is not detected in libsnark on the prover side.

Setting all the pk equal to the identity, except for one wire: This is a practical attack. Similar to the attack above, the malicious verifier will set all the $pk_A, pk'_A, pk_C, pk'_C \in G_1$ equal to 0. Consequently, the proof π will reveal the value of j as above.

7.8.3   Open Problems for zk-SNARK

To instantiate the open problems for zk-SNARK, we follow the Xie et al. [237]'s work to generalize following two major open problems for implementing zk-SNARK in practice.

Improving verification time. According to [237], Libra's verification time is already fast in practice compared to other systems. Most of the time, as observed, is spent on the zero-knowledge verifiable polynomial delegation scheme (zkVPD) protocols using bilinear pairings. Although the masking polynomials are small, the verification of zkVPD still requires $\mathcal{O}(s_i)$ pairings per layer. A potential solution is to replace the pairing-based zkVPD in Libra [237] with any of the zero-knowledge proof systems we compare with as a black-box.

Removing trusted setup. Replacing the pairing-based zkVPD in Libra [237] with other systems without trusted setup may affect the succinctness of the verification time on structured circuits. Designing an efficient zkVPD protocol with logarithmic proof size and the verification time without a trusted setup is challenging.

7.9   Chapter Summary

Zero-knowledge proof (ZKP) has long been used to achieve zero leakage of private information. It is also a very useful tool to achieve privacy-preservation for blockchain. In fact, some variants (like zk-SNARK) have been tentatively used in crypto-currencies. This chapter dives deeply into the study of zero knowledge proof system. It first reviews background and knowledge of zero-knowledge. Specifically, it shows sequential instructions to build zk-SNARKs in the generic method. This chapter then reviews three notable works of zero knowledge proof: GS08, GR16 and CA07 schemes. The review of the above schemes help readers gain empirical knowledge of designing and analyzing ZKP schemes. Finally, this chapter reviews the practical implementation of zk-SNARK in Zcash, its performance and other associated problems. After reading this chapter, the reader should be able to: (1) Understand the meaning and definition of ZKP. (2) Know a generic method to construct zk-SNARKs. (3) Learn basic principle to design and analyze ZKP schemes. This chapter is summarized as follows: (1) It is not trivial to generate a zk-SNARK. The reader should devote hard works in learning what zk-NARKs is about and how it works. (2) Current ZKP schemes develop fast and extend to various research fields. The analysis of ZKP scheme generally does not involve complex or skilled proof techniques. However, to fully figure out the meanings of many notations and terms, the reader ought to follow specific research line and keep track of relevant works. (3) Due to complex and sophisticated design, it is generally infeasible to derive practical zk-SNARK. Meanwhile, the assumption of honest setup or entity to initiate the system parameter contradicts the blockchain's decentralization. Therefore, a true ZKP scheme or variant (like zk-SNARK) designed for blockchain should overcome the above problems.

Chapter 8

Tools as Optimizations for Blockchain

8.1 Chapter Introduction

This chapter identifies several significant problems hindering the successful and long-term development of the blockchain, namely, security, privacy, efficiency, scalability, inflexibility and vulnerability. Following the categorized issues, we enumerate a useful tool and idea to improve the blockchain. These topics provide unique perspectives for the reader to better pursuit studies of the blockchain.

After reading this chapter, you will be able to:

- Know the basic definition of a given tool or primitive discussed in this chapter.

- Know the major problems associated with the blockchain.

- Know what the possible solutions to a given problem are.

8.2 Main Problems in Blockchain

In this section, we review four major problems of the blockchain. These problems can not be easily addressed in a short time or in a trivial manner.

8.2.1 Security Problem

The blockchain utilizes asymmetric cryptography, such as public-key encryption, signature, hashing and zero-knowledge for identification, authentication, authorizing the transactions, etc. Only those who possess the corresponding private key can access the private wallet and perform the secret operation. Specific security services and infrastructures guarantee these behaviors. In other words, presumably, the private key is the only security instrument that authorizes the lawful owner. Therefore, the leakage of the private key will result in an instant break down of the entire system.

Unfortunately for those ambitious hackers, blockchains are decentralized and distributed across peer-to-peer networkscontinually updated and kept in sync. Because they are not contained in a central location, blockchains do not have a single point of failure [259] and cannot be changed from a single computer. It would require massive amounts of computing power to access every instance (or at least a 51 per cent or the majority) of a particular blockchain and alter them all at the same time. There has been some debate about whether this means smaller blockchain networks could be vulnerable to attack, but a verdict has not been reached. In any case, the bigger your network is, the more tamper-resistant your blockchain will be.

Although public-key cryptography used in the blockchain is considered one of the best and strongest cryptographic methods available, there is no additional safety net to protect the blockchain users from losing or unwillingly sharing their private key with others. This leads to our next topic: privacy issue.

To summarize, the private key of a blockchain account is an important target to protect. Some people MAY consider the lack of additional security measures a limiting factor for blockchain usage.

8.2.2   Privacy Problem

The blockchain is a purely distributed peer-to-peer ledger that maintains the whole history of transaction data. All transaction details such as the goods and the amount being transferred, the involved accounts, and the time of transfer are accessible to everyone. This public verifiability, though, allows every peer to clarify ownership and verify new transactions and contributes to the disruption from loss of sensitive information.

Privacy refers to sensitive informationconcerning a person (or company), such as personal financial state and real identity behind pseudonymity. Since blockchain is increasingly involved in people's social lives, some privacy hidden among the ever-increasing data recorded on the chain. Due to the easily-accessible network and powerful analysis technique, privacy can be easily breached and extracted. Identity and transaction privacy are the two major privacy involved in the blockchain. Following levels of leakage can summarize the threats against the above two privacy:

Network-Layer Privacy Threat: Revelation of users' real identity from pseudonymity is considered a significant privacy leakage at this stage. The attacker can eavesdrop or actively collect communicated data between neighbor nodes via network techniques.

This allows him to analyze the topology path to locate the original sending node of a transaction. Suppose senders' and recipients' addresses are publicly recorded (e.g. like Bitcoin). Once locating the original sending node, the attacker can thus link the associated pseudonymity with its real identity.

Transaction-Layer Privacy Threat: Extracting user's financial information is a major threat at this stage. For Bitcoin, the transaction is publicly verifiable. The attacker can analyze the transaction log to derive the user's financial information, such as the balance of the personal account, currency flow, trade pattern, etc. Besides, identity privacy is also at the risk of exposure. The attacker can deduce the user's real identity from pseudonymity by analyzing associated transactions. For example, based on the empirical knowledge that multiple outputs in one transaction mostly indicates an identical set of users (e.g. multiple miners assigned to the same mining pool). For another example, an account frequently receiving large amounts of Bitcoins with zero output, is supposed to be a saving account for a company or a corporate.

Application-Level Privacy-Leakage: Users' misbehaviors in operating blockchain services results in privacy leakage. For instance, user adopts unique pseudonymity for each website. This provides the attacker with ample chances to successfully guess his real identity. Meanwhile, the type of application or infrastructure under which the blockchain is implemented provides attackers with particular approaches to derive privacy. For instance, regarding smart grid-based blockchain, the attacker can extract user' power consumption data from the blockchain log to guess his financial capabilities, living habits, absence hours from home, etc.

8.2.3 Efficiency and Scalability Problem

The blockchain utilizes an immutable append-only data structure that requires the solution of a hash puzzle every time a new block is added. Solving that hash puzzle is time-consuming on purpose. So, the hash puzzle solution makes it costly to manipulate the history of transaction data. Unfortunately, this security measure comes at the price of reduced processing speed and hence limited scalability. Solving the hash puzzle or providing the proof of work is computationally expensive on purpose. The proof of work is expensive, and therefore, the whole blockchain incurs costs. The magnitude of these costs depends on the difficulty of the hash puzzles. As Bitcoin becomes more and more popular, it becomes more time-consuming to confirm a transaction. Since the number of transactions has increased in the network, another blockchain scalability problem arises: the time-consuming process of transaction execution. To summarize, there are several serious problems affecting blockchain efficiency and scalability, including block size, response time and fees.

8.2.4 Inflexibility and Vulnerabilities

The blockchain is a complex technical construction consisting of a variety of concepts and protocols optimized and adapted to one another. Changing that fine-tuned ecosystem can be very challenging. When applying blockchain as a trust-layer for Internet of Things (IoT) or big data, security and privacy rules are applied [260]. However, as recently reported by Ali et al. [42], the newly launched blockchains suffer from powerful attacks (such as 51% attack [1]) due to small computing pools. Malicious users can easily manipulate the blockchain with considerable but not unimaginably-large computing resources. As a fact, manipulation over 60% of the network power has been witnessed [42], which is a severe threat against the blockchain. At this point, Ateniese et al. [215] introduced the redactable blockchain to efficiently remove illegal and malicious contents from the blockchain without causing inconsistency of the block hash. As witnessed by the catastrophic loss caused by the "DAO" incident [261] to one of the dominant crypto-currencies induced by hackers, it is necessary to grant newly started blockchains with the power of redaction to be immunized against such attack. However, there is no established procedure for changing or upgrading major components of a blockchain once it has started its operation. Despite various extensions (consortium or private blockchain), public blockchain is generally known as an immutable (or uneditable) trust-layer. Immutability can be explained as follows:

Block Hash Immutability: By using cryptographic hash functions, a hash value is computed and stored in each block header which links to the previous block. Due to the intractability of finding a hash collision (aka collision resistance), it is hard to forge another hash value which outputs the same hash value.

Transaction Immutability: To deal with a transaction, the sender and receiver should be part of a public-key infrastructure. They follow the digital signature scheme for authentication as negotiated. Due to the unforgeability of the digital signature, it is hard to forge a signature and pass the verification for a transaction to which the signature is committed. Thus, transactional immutability is achieved.

History Immutability: For public blockchain, the remote history is immutable since it is both technically and economically infeasible to reverse (as discussed by Nakamoto). For the recent history (e.g. the next 5 or 6 blocks), although some minor forks may occur (by rare chance), they will be eventually discarded once most miners keep following a longer chain.

Due to the anonymous, freely-accessible natures of networks and lack of financial surveillance, early blockchains are exploited by malicious users. As early discussed in Section 3.3.1, recent years have witnessed various vulnerabilities in blockchains. They can be categorized by vulnerabilities from the data layer, network layer, consensus layer, incentive layer, contract layer, and the application layer. Specifically, the concrete attack targeting specific blockchain layer and Altcoin is listed in Table 3.4 in Section 3.3.1.

8.3   Tools to Enhance Security

In this section, we review some security primitives utilized as basic building blocks for enhancing security of blockchain.

8.3.1 Commitment

A commitment scheme is an interactive or non-interactive protocol between two parties, the sender S holding a message, and the receiver R. A commitment scheme is a tuple (Gen, Commit) such that on input security parameter 1^λ, Gen outputs some system parameters Param for the commitment scheme. Commit is a determinist algorithm that on input Param, a value x and a random number r, outputs $C = $ Commit(Param, x, r). C is called the commitment of x, and r is sometimes called an *opening*. Following definitions given by Man Ho Allen in ref. [262]. We give two basic security requirements for commitment scheme as follows.

Definition 8.1 *(Perfect Hiding). A commitment scheme* (Gen, Commit) *is secure if it holds the following two properties:*
(1) The adversary chooses x_0 and x_1 and receives a commitment at random. (2) The adversary cannot efficiently distinguish between the origins of the committed value.

$$\Pr\left[\begin{array}{c} \text{param} \leftarrow \text{Gen}(1^\lambda); (x_0, x_1) \leftarrow \mathcal{A}(\text{param}); \\ b \in_R \{0,1\}; r \in_R \{0,1\}^\lambda; \\ b' = b; \\ C = \text{Commit}(\text{param}, x_b; r); b' \leftarrow \mathcal{A}(C); \end{array}\right] \leq 1/2 + \text{negl}(\lambda). \quad (8.1)$$

Definition 8.2 *(Binding). No efficient adversary $\mathcal{A}$ can open a commitment in two different ways. Specifically,*

$$\Pr\left[\begin{array}{c} \text{param} \leftarrow \text{Gen}(^\lambda); (x_0, x_1, r_0, r_1) \leftarrow \mathcal{A}(\text{param}) : \\ x_0 \neq x_1 \wedge \text{Commit}(\text{param}, x_0; r_0) = \text{Commit}(\text{param}, x_1; r_1) \end{array}\right] \leq \text{negl}(\lambda). \quad (8.2)$$

Pedersen commitment [155] is a famous commitment scheme.

8.3.2 Trapdoor Commitment

Trapdoor commitment is a special class of commitment schemes useful in many cryptographic systems. As we systematically reviewed in Chapter 6, chameleon hash (or PKH) is a simple version of trapdoor commitment. One important field of applications of trapdoor commitments are zero-knowledge proofs. One formalizes trapdoor commitment as follows.

Definition 8.3 (Trapdoor Commitment Scheme) *A trapdoor commitment scheme* TC *is a tuple of algorithms (* TC.Key, TC.Com, TC.Vrf, TC.Sim, TC.Open) *such that:*

TC.Key$(ck, tk) \leftarrow (1^\lambda)$: This algorithm takes security parameter 1^λ. It generates a commitment key ck and a secret trapdoor key tk.

TC.Com$(c, d) \leftarrow (ck, m)$: This algorithm on input m and ck, outputs (c, d), where c is a commitment of the input message m and d is the corresponding decommitment value.

TC.Vrf$(0, 1) \leftarrow (ck, c, m, d)$: This algorithm verifies whether the commitment c corresponds to the message m with an opening d with respect to ck. If so, returns 1; otherwise, 0.

TC.Sim$(c, ek) \leftarrow (ck, tk)$: This algorithm takes the trapdoor key and commitment key. It computes a equivocal commitment c and a corresponding equivocation key ek.

TC.Open$(d) \leftarrow (ck, c, m, ek)$: This algorithm takes an equivocal commitment c, a message m, a commitment key ck, and an equivocation key ek. It outputs a decommitment d for which TC.Vrf$(ck, c, m, d) = 1$.

Trapdoor commitment schemes have been used to construct zero-knowledge proofs [263] in which the prover and verifier exchange only a constant number of messages [264–266], concurrent zero-knowledge protocols where the verifier talks to several instances of the prover in parallel [267,268] and resettable zero-knowledge [269] where the verifier is even allowed to reset the prover to some previous step of the protocol. Additionally, trapdoor commitments play a vital role in constructing secure signature schemes without relying on the strong random oracle assumption [270,271]. Also, they turn out to be quite useful for the construction of secure undeniable signatures [218].

8.3.3 Merkle Hash Tree

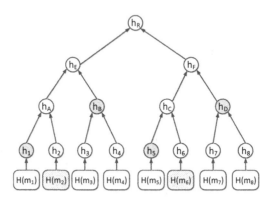

FIGURE 8.1: Merkle Hash Tree

Merkle Hash Tree (MHT) is a binary tree for verifying data integrity. It works by inputting the hash of each file block as the leaf node and iteratively computing the parent node by two child nodes until deriving the Merkle root h_R. As given in Figure 8.1, we instantiate how to compute the Merkle root of file consisting of 8 blocks. First, inputting the hash of each block as leaf node, i.e. $h_i = H(m_i)$ for $1 \leq i \leq n$. Here, H denotes a collision-resistant hash function. The verifier only needs to keep h_R at local for the verification. Suppose the verifier issues $\{m_2, m_6\}$ as a challenge. The prover should return both the challenged nodes, brother nodes $\{h_1, h_5, m_2, m_6\}$ (nodes in grey) as well as the corresponding sibling path $\{h_B, h_D\}$ (nodes in full line) as a proof $prf_{mht} = \{m_2, m_6, h_1, h_5, h_B, h_D\}$. On receiving the proof prf_{mht}, the verifier can compute $h_2 = H(m_2)$, $h_6 = H(m_6)$, $h_A = H(h_1||h_2)$, $h_C = H(h_5||h_6)$, $h_E = H(h_A||h_B)$, $h_F = H(h_C||h_D)$, and $h'_R = H(h_E||h_F)$. Then, the verifier can compare the derived value h'_R with the valid one h_R to determine the validity.

8.3.4 Public-Key Encryption

Public-key encryption is a crucial tool in public-key cryptography, which has demonstrated many useful applications such as data confidentiality, key exchange, oblivious transfer, etc. We give systematic and hieratical analysis on the public-key encryption schemes used or considered as eligible to be used in blockchain in Section 5. To design secure PKE for the blockchain, academia accepted that IND-CCA2 is a practical security level for a proposed PKE to achieve. This chapter leverages the practical security and proof techniques to achieve practical design and analysis of PKE scheme.

8.4 Tools to Enhance Privacy-Preservation

In this section, we review several cryptographic primitives as solutions to achieve or strengthen privacy-preservation in blockchains.

8.4.1 Zero-Knowledge Proof

Goldwasser et al. [55] formally proposed notion of zero-knowledge proof. Zero-Knowledge Proof (ZKP) protocol allows a prover to convince a verifier of knowing a secret via a statement while the verifier learns nothing about the secret. Non-Interactive Zero-Knowledge Proof (NIZK) is a variant of ZKP where proof can be independently verified by anyone once generated without interactions. The Zero-Knowledge Succinct Non-Interactive Argument of Knowledge (zk-SNARK) is another practical variant of ZKP. It produces argument with small and constant size while formalizing security by assuming adversary with bounded computational power. Here, the difference between argument and proof relates to the assumption of adversary's power. ZKP and its variants are building blocks for many cryptographic paradigms, such as privacy-preserving encryption, signing, etc. Mainstream ZKP variants like zk-SNARK suffer from centralized key generation. It relies on an honest generator's assumption to output the common reference string (a parameter outputted from the key generation). Although this problem has been partially solved by introducing multi-party computation, computing expenditures rise significantly with the number of participants involved. Meanwhile, solving this scalability problem is not trivial. In chapter 7, we systematically studied ZKP and relevant knowledge.

8.4.2   Group Signature, Ring Signature, and Variants

Anonymous signature generally refers to a primitive, which allows a signer to authenticate a message without exposing his identity to the verifier. It generally implies a group signature or ring signature. Chaum and Van Heyst [75] proposed the group signature, which allows a group manager to set up a group of users so that any group members can sign on behalf of the whole group. Rivest et al. proposed the notion of ring signature, which eliminates the need for a group manager that greatly simplifies the setup phase for signing. Unlike the group signature, the signer can randomly choose a set of users' public keys to form a ring in which his public key is hidden. Since such a ring is formed spontaneously, other users are unaware that they are summoned for generating the signature. As no central trust is required, it achieves unconditional anonymity. However, it also raises the problem of tracing dispute signature since no central trust can be counted on. As a partial solution, linkable ring signature [77] is proposed to detect ring signatures produced by the same signer through linkability of tags via computations. This could be an ideal primitive to detect double-spending in cryptocurrencies. Also, designing constant and fixed signature size is a challenge. In chapter 4, we conduct a systematic review and detailed analysis on this primitive.

8.4.3   Multi-Party Computation (SMPC)

Andrew Yao [78] proposed secure two-party computation and Goldreich et al. generalized it to multi-party computation. MPC protocol allows multiple parties to jointly compute a value over their private inputs while the adversary learns nothing about each secret input but the output.

SMPC is a protocol to ensure secure and complicated operations between trustless parties. Secret-sharing-based MPC (SS-MPC) relies on oblivious transfer (OT) to obscure the participant's view of the protocol's ending cycle. As a result, it is for each participant to cheat and recover others' privacy during collaborative computations. According to the different assumptions of colluded parties and how they collude (statically or adaptively), SMPC protocols vary in its design and security. So far, MPC has been successfully implemented in achieving e-lottery, e-voting, e-auction, etc. It has been actively studied and used to design cryptocurrencies [80] and other projects (e.g. Enigma [81]). However, it suffers from scalability and inefficiency problems when introduced in the blockchain. Since blockchain is free to join and attracts many users, complexity costs generated from SMPC rise dramatically when the number of users surge.

8.4.4 Fully Homomorphic Encryption

Homomorphic encryption (HE) is a particular type of encryption which allows a third party to manipulate encrypted data without decryption. Since direct computation is permitted on the encrypted data without violating the decryption, both data operability and privacy are guaranteed. HE can be interpreted as a non-interactive version of secure multi-party computation since no one is required to collaborate to generate the output. Rivest proposed the first fully homomorphic encryption (FHE) in 1978, which supports both addition and multiplication operations on encrypted data. It was attractive as it was still after 30 years ago that Craig Gentry [83] proposed the first plausible and practical FHE. Despite ongoing studies, current FHE schemes still can not support the real-time applications. To tackle, optimizing works start from various perspectives, including software, hardware and schematic optimization. Besides, even though proposed FHE satisfy specific security goals, it is suggested to achieve IND-CCA1 security for general purpose applications.

8.5 Tools to Improve Flexibility and Vulnerabilities

In this section, we review several tools as solutions to the inflexibility and vulnerabilities of blockchains.

8.5.1 Chameleon Hash as Solution

This type of work focuses on the cryptographic study of the chameleon hash function. Ateniese et al. [215] formally proposed to use the chameleon hash as the core element to drive the redactable blockchain. Ashritha et al. [272] proposed to split the trapdoor key (private key) among multiple validators and reconstruct it via multi-party computation for secure redaction. As a result, redaction is safely performed among multiple entities without mutual trust.

Informally, the chameleon hash function can be comprehended as a trapdoor one-way hash function where it is hard to find a hash collision if the trapdoor is unknown. Krawczyk and Rabin [218] first proposed chameleon signature in 2000. Later on, Ateniese et al. [219] improved it by an extension of identity-based setting. Also, they identified and solved the key-exposure problem, where it asks that it is infeasible to extract trapdoor key from any given collisions. We formalize chameleon hash as public key hashing scheme in Chapter 6 with systematic review and analysis. To exemplify the use of chameleon hash as a solution to improve blockchain's inflexibility (by transforming immutable blockchain into mutable one), we review relevant works in Figure 8.1 in terms of various functions.

As it is shown in Table 8.1, our proposed RCH (as discussed in Section 6.6) inherits more merits than other works. Meanwhile, most hash schemes do not rely on pairing-based computations. However, by considering the versatile functions of pairing-based infrastructure and the strong computational power owned by personal computers nowadays, it is reasonable to involve pairing computations. In this work, we specifically show how to introduce ephemeral trapdoor so that forgery of a hash collision is allowed for the public (while assuming ephemeral trapdoor is given). To prevent trapdoor abuse, we enforce the periodical expiration of the generated hash and the ephemeral trapdoor. This is a useful tool to derive blockchain redaction securely and intelligently. To our knowledge, our work is the first revocable chameleon hash scheme.

TABLE 8.1: Comparison of Recent Chameleon Hash for Blockchain's Redaction

Scheme	RCH	TCH [150]	PCH [273]
Decentralization	No	Yes	No
Public Redaction	Yes	No	No
Hash Revocation	Yes	No	No
Identity-based	Yes	No	No
Fine-grained Redaction	No	No	Yes
Assumptions	CDHP	CDHP	DLP

8.5.2 Malleable Signature as Solution

8.5.2.1 Malleable Signature

A signature scheme is malleable if, on input a message and a signature, it is possible to efficiently compute a signature on a related message, for a transformation that is allowed on this signature scheme [274]. Malleable signatures allow the signer to control alterations to a signed document. The signer limits alterations to certain parties and certain parts defined during signature generation. Admissible alterations do not invalidate the signature and do not involve the signer. These properties make them a versatile tool for several application domains, like e-business and healthcare [275].

8.5.2.2   Chameleon Signature

Krawczyk and Rabin [85] first proposed Chameleon Signature (CS) to achieve the notion of "undeniable signature" [276]. Later, Ateniese and Medeiros [219] extended [85]'s scheme by proposing the first identity based CH against the key-exposure problem. In [219], a customized identity is used to limit the leakage of the trapdoor to every single transaction. Chen et al. [149] proposed the first key-exposure free CH in the GDH groups. Ateniese and Medeiros [67] presented three key-exposure CH schemes. Later, Chen et al. [148] proposed a key-exposure free CH and CS based on the DLP. Recently, Camenisch et al. [86] proposed CH with ephemeral trapdoors to prevent the trapdoor holder from finding collisions if no other ephemeral trapdoor is leaked. Krenn et al. [220] also proposed chameleon-hashes with dual long-term trapdoors where the hashing party can choose between using a fresh second trapdoor and reusing an existing one. So far, no TCH is proposed.

8.5.2.3   Sanitizable Signature

Ateniese et al. [277] proposed the first Sanitizable Signature (SS) based on CS [85] which allows a semi-trusted third party to modify signatures moderately. Later on, SS schemes follow inner-and-outer signature structures to avoid the linear cost of sanitization and adopt different underlying signatures to offer distinct properties [278]. Brzuska et al. [278] formally investigated several security properties and their relationships. Then, Brzuska et al. [279] defined unlinkability specifically to prevent linking sanitized signatures to the same message. Canard et al. [280] considered a multiplayer (n signers and m sanitizers) scenario based on the work by Brzuska et al. [279].

8.5.2.4   Redactable Signature

A Redactable Signature Scheme (RSS) for a string of objects supports the verification even if multiple sub-strings are removed from the original string. It is crucial that the redacted string and its signature do not reveal anything about the removed sub-strings' content. Unlike the usual signature schemes, the RSS has additional requirements on privacy, e.g. information of the removed strings should be hidden. Redactions are especially useful if the original signer is not reachable any more, e.g. in case of death or if it produces too much overhead to re-sign a message every time an alteration is necessary. RSSs allow removing blocks from signed data. State-of-the-art schemes have public redactions, i.e. any party can remove parts from a signed message [281,282].

8.5.2.5   Redactable Blockchain

In recent years, various researches have been proposed to achieve the redactable blockchain [215] (or immutable blockchain [283]).This topic has quickly extended to various aspects which generally seek to achieve flexible and re-writable transaction history. We briefly review relevant works as follows.

Redactable blockchain indicates the blockchain with an editable (redactable) block or revocable transaction for security or privacy concerns. As shown in Table 8.2, various schemes were proposed to achieve this goal.

Reversecoin [87] is the first cryptocurrency to support transaction revocation. It allows the user to revert a transaction in a configurable timeout by using off-line keys. Although this project did not succeed, it has offered valuable insights in designing redactable blockchain.

TABLE 8.2: Overview of Redactable Blockchain Works

Scheme	Technique & Principle	Pros&Cons
Reversecoin [87]	uses offline key to reverse transaction in a timeout period	the first revocable coin, but did not succeed
Redactable Blockchain [215]	proposes the notion of redactable blockchain formally with chameleon hash	allows to rewrite block history, but many issues are not discussed
Thwarting Insertion [283]	proposes a content detector with minimum fees to reject malicious contents on chain	gives conceptual suggestions, but does not provide security analysis
μchain [284]	proposes μchain to enable changes of transaction state to erase relevant blockchain records	does not rely on complex cryptographic schemes, but introduces high computing costs
Permissionless [216]	proposes voting-based consensus protocol to achieve redactable blockchain	relies on complex cryptographic protocol, but it is fairly efficient to run
Enhanced Redactable Blockchain [272]	uses multi-party computation to reconstruct trapdoor for secure redaction among multiple validators	enhances redaction security at the expense of introducing complex interactions and computing

Other related works are [150,224], each scheme is designed for specific application scenario.

Ateniese et al. [215] proposed the notion of redactable blockchain in EuroS&P 2017. They introduced chameleon hash as the core element to derive the redactable blockchain so that block contents could be re-written without causing any major hard forks. Despite their efforts, there are still some unresolved problems. For example, how to reach an agreement on redaction or how the chameleon hash is used during redaction.

Then, Matzutt et al. [283] proposed a content detector with minimum fees to reject malicious contents insertion on chain, but no security analysis is provided to validate their proposal. Also, Puddu et al. [284] proposed μchain to alter transaction states to change blockchain records, but their proposal induced high computing costs. Later, Deuber et al. [216] proposed an efficient and redactable blockchain by consensus-based voting and policy to dictate redaction. Only relatively low overhead is yielded for their proposal in comparison with the immutable one. Recently, Ashritha et al. also [272] proposed to split the trapdoor key (private key) among multiple validators and reconstruct it via multi-party computation for secure redaction.

To summarize, Politou et al. [216] proposed a review article to list challenges and developments for the redactable blockchain. As discussed in their work, solutions to redactable blockchain were not limited to cryptographic methods. Solutions extended to a consensus protocol, malware detection, adjustment of transaction fees, etc. The philosophy of the above solutions is to let major nodes forget about the problematic block history. The mutability of the blockchain is supposed to have positive effects on industry and academia; it answers how to achieve "the right to be forgotten" to erase harmful and sensitive information from the blockchain.

8.6 Tools to Improve Efficiency and Scalability

In this section, we review several tools as potential solutions to improve the efficiency and scalability of the blockchain.

8.6.1 Accumulator

Benaloh and De Mare [230] formally introduced accumulator. It allows the representation of a set of elements by a single value of size independent of the set's cardinality. Accumulators are mainly used as building blocks for more complex cryptographic systems. For instance, they are used to construct anonymous identification in ad hoc groups [285], and anonymous credentials with efficient revocation of credentials [286].

We briefly explain how accumulator works. Given a value v, one can accumulate an element y into v by invoking the accumulating function f as $v \leftarrow (v, y)$. Therefore, if u is an initial value, $v = \mathsf{f}(\cdots \mathsf{f}(\mathsf{f}(u, y_1), y_2) \cdots, y_n)$ represents a value v into which elements $y_1, y_2, \cdots, y_n$ have been accumulated. Furthermore, for any element y and any value v, there exists a witness w for y w.r.t. v if and only if y has been accumulated into v. One can thus, by demonstrating (the knowledge of) a valid corresponding witness, prove that an element has been accumulated into a value. An accumulator is required to satisfy properties including efficient generation, quasi-commutativity, efficient evaluation and membership witnesses. For more details, refer to ref. [286,287].

8.6.2   Bloom Filter

A Bloom Filter (BF) [288] is a simple space-efficient randomized data structure for representing a set in order to support membership queries. It is a probabilistic data structure for the approximate set membership problem. It allows a succinct representation T of a set $\mathcal{S}$ of elements from a large universe $\mathcal{U}$. The space efficiency is achieved at the cost of a small probability of false positives, but often this is a convenient trade-off. Bloom filters' main advantage is their memory and time efficiency, since they require less space than other data structures to store elements in the set and less time to perform membership queries.

For elements, $s \in \mathcal{S}$ a query to the BF always answers 1. Ideally, a BF would always return 0 for elements $s \notin \mathcal{S}$, but the succinctness of the BF comes at the cost that for any query to $s \notin \mathcal{S}$ the answer can be 1, too, but only with a small probability (called the false-positive probability). The space efficiency is achieved at the cost of a small probability of false positives, but often this is a convenient trade-off. Although bloom filters were invented in the 1970's [288], and it was until recent years that bloom filters have been heavily used in applications (see e.g. [289,290]).

Next, we review the definition for the original construction of bloom filters by Bloom [291]. Refer to [291] for more details.

Definition 8.4 *A bloom filter B for set $\mathcal{U}$ consists of algorithms B=(BFGen, BFUpdate, BFCheck), defined as follows.*

BFGen(m, k): *This algorithm takes as input two integers $m, k \in \mathbb{N}$. It samples $H_1, \cdots, H_k$, where $H_j : \mathcal{U} \to [m]$, defines $H := (H_j)_{j \in [k]}$ and $T := 0^m$, and outputs (H, T).*

BFUpdate(H, T, u): *Given $H = (H_j)_{j \in [k]}$, $T \in \{0,1\}^m$, and $u \in \mathcal{U}$, it defines $T' := T$. Then, setting $T'[i]$ to denote the i-th bit of T', it sets $T'[H_j(u)] := 1$ for all $j \in [k]$, and returns T'.*

BFCheck(H, T, u): *Given $H = (H_j)_{j \in [k]}$, $T \in \{0,1\}^m$ where we write $T[i]$ to denote the i-th bit of T, and $u \in \mathcal{U}$, it returns a bit $b := \wedge_{j \in [k]} T[H_j(u)]$.*

8.6.3   Micro Payment

Micropayments [292] are increasingly being adopted by a large number of applications. This paradigm allows participants to exchange monetary incentives at a small scale, e.g. pay per minute in online games. By aggregating these small transactions into a few larger ones, and using cryptocurrencies, today's decentralized probabilistic micropayment schemes can reduce these fees. It is considered as an ideal solution to the scalability problem of blockchain. However, processing micropayments individually can be expensive, with transaction fees exceeding the payment value itself. For example, assuming the average base cost of a debit or credit card transaction in the US is around 21 to 24 cents, and 23 to 42 cents [293], respectively. In cryptocurrencies, such a fee could be even higher, e.g. above \$1 in Bitcoin.

Alternatively, probabilistic micropayment schemes have emerged as a solution that fits the criteria outlined above [294]. In these models, the required payments are locked in an escrow and micropayments are issued as lottery tickets. In short words, each payment is probabilistic. To explain, each ticket has a probability p of winning a lottery, and when it wins, produces a transaction of β currency units. As cryptocurrencies evolved, several initiatives have attempted to convert these schemes to distributed ones [295]. This is done by replacing the trusted party with the miners and utilizing the blockchain to provide public verifiability of the system operation. However, various drawbacks are found in these works, including:

1. Against the customer's will to use the payment mandatorily.

2. Require a merchant to report the lottery outcome to a customer.

3. Result in a large number of escrows.

4. Rely on computationally-heavy cryptographic primitives [295].

In general, payment channels and networks suffer from the high collateral cost of setting up multiple escrows when constructing payment paths between transacting parties. These leads to centralized trust and other issues [296]. Besides, each hub on the path charges a fee to relay payments. With micropayments, such a setup would be infeasible because these fees could be much larger than the payments themselves. It is still challenging to devise practical a micropayment scheme which satisfies massive processing requirements.

8.6.4 Lightning Payment Channel

The Bitcoin [1] blockchain holds great promise for distributed ledgers, but can not cover the world's commerce anytime shortly. The payment network Visa achieved 47,000 peak transactions per second (tps) on its network during the 2013 holidays [297]. Achieving Visa-like capacity on the Bitcoin network is not feasible today. No home computer in the world can operate with that kind of bandwidth and storage. To achieve much higher than 47,000 transactions per second, Bitcoin requires conducting the Bitcoin blockchain transactions. To alleviate scalability issues, the cryptocurrency community is continuously inventing new protocols and technologies. Lightning Network (LN) [298] is designed to amend the scalability and privacy issues of Bitcoin. It is a payment channel network aiming to settle transactions faster, cheaper, and more private. It provides possibilities for the blockchain to scale at a considerable level. A significant research line is focused on introducing the notion of off-chain transactions [299].

To evaluate LN efficiency and profitability, Miller et al. [299] designed a traffic simulator to analyze the routing costs and potential revenue at different nodes. They also designed a method to predict the optimal fee pricing policy for individual nodes in the cheapest path routing. After the launch of

LN [299], several studies have investigated the graph properties of LN [300]. To conclude, the use of LN generally supports millions to billions of transactions per second across the network. By transacting and settling off-blockchain, the Lightning Network allows for exceptionally low fees. However, due to the use of bidirectional payment channels between two nodes which combine created smart contracts, the channel will close and be settled on the blockchain if at any time either party drops out of the channel. Due to the nature of the Lightning Network's dispute mechanism, it requires central trust on a watchtower to monitor fraud. How to erase the central trust from the current micropayment schemes is a challenging issue.

8.7   Chapter Summary

This chapter identifies several significant problems hindering the successful and long-term development of blockchain, namely, security, privacy, efficiency, scalability, inflexibility and vulnerability. Following the categorized issues, we enumerate a useful tool and idea to improve blockchain. These topics provide unique perspectives for the reader to better pursuit studies of blockchain. After reading this chapter, the reader should be able to: (1). Know the basic definition of a given tool or primitive discussed in this chapter. (2). Know the significant problems associated with blockchain. (3). Know what the possible solutions to a given problem are. To summarize, cryptographic researches are increasingly affected by the studies on the blockchain, resulting in an overlapping research zone with hot spots (e.g. ring signature, zk-SNARK, etc). The tool and cryptographic primitives enumerated in this chapter provide unique perspectives for the reader to better pursue the blockchain studies. It also gives a beginner a good starting point to follow and carry out their researches accordingly.

Chapter 9

Regulation and Economies of Blockchain

9.1 Chapter Introduction

This chapter is devoted to investigating global development in regulations and flexible and practical countermeasures to achieve regulatory goals. It categorizes and reviews business activities closely related to the blockchain. This is useful for the researcher to learn global developments in the blockchain regulations. It also helps the practitioner learn the business patterns and potential weak links in conducting the blockchain regulatory affairs.

After reading this chapter, you will be able to:

- Know global developments in the blockchain regulation. Know what is the regulatory sandbox.

- Know challenges and possible solutions for the blockchain regulation.

- Know business events and affairs associated with the blockchain services.

9.2 Background

In recent years, the fast development and massive implementation of blockchain technology have caught attentions worldwide. However, many problems have arisen during this development, such as security, efficiency, scalability, etc. Among these problems, financial and security problems are the most needed to address. The main reason is simple: blockchain is first and foremost, the core idea for the most successful digital currency (Bitcoin [1]). Throughout the past 12 years, from early reports on Bitcoin regarding "Darknet trade [301]", "financing terrors [302]", to recent incidents like "ICO bubbles [303]", the abuse of crypto-currency disrupts finance order. Meanwhile, blockchain suffers from attacks in various aspects (system-level [304], software or hardware-level [305], protocol-level [306], etc.). "THE DAO" incident [63] is one of the most notorious incidents caused by a group of trained hackers in 2017. They take advantage of the smart contract's loophole, which causes significant loss to the Ethereum [2]. Another example is the widespread computer virus "WannaCry" [307], which leads to billion-dollar loss to computer users worldwide. "WannaCry" exploits the anonymity and untraceability of Bitcoin to retrieve its illegal gains from blackmail. Incidents like the above severely affect the financial order as well as cyber security of every nation in the world connected to the Internet.

9.3   Blockchain Regulation

In this section, we review the background, challenges, tools to perform blockchain regulations.

9.3.1   Global Developments of Regulation on Blockchain

In recent years, many governments and financial sectors have released relevant statements and legislations to regulate the blockchain and other financial technologies (FinTech [308]). We list the development of the blockchain regulation in the five major economies during 2013–2019 in Table 9.1. As demonstrated in Table 9.1, all listed nations acknowledge and encourage blockchain-based business and agree to enforce strict supervision and regulation on the blockchain industry. For ease of analysis, we categorize the years of regulation by three different attitudes:

- **Diverse (2008–2013):** At this stage, many countries reflect either diverse or silent on blockchain regulation. Specifically, China notice to fend off Bitcoin risks while Germany shows a positive attitude by acknowledging Bitcoin in the first place [309].

- **Cautious (2014–2018):** During this time, all listed countries show careful attitude towards the blockchain regulation. They frequently release regulation statements as well as legislation. However, they also actively promote blockchain development in order to boost its economy. Their regulation principle is per their specific development's objectives as well as a national strategy. For example, the Chinese government runs regulation via multi-department co-operation and includes it in its 13th Five-Year Plan [310].

- **Tolerable (2019):** With the successful implementation of regulatory sandbox [311] as well as in-depth studies on blockchain, many countries have developed tolerable attitude towards blockchain. On the one hand, they continue to supervise and regulate the blockchain industry. On the other hand, they seek to stimulate the blockchain industry at maximum to boost their economies. A bad example is German regulation on blockchain [312], which fails to stimulate its national blockchain industry, in turn, receives little economic revenue from blockchain development in comparison with its counterparties (like Switzerland).

TABLE 9.1: Global development of blockchain regulation

Year	USA	China	Japan	Germany	UK
2008-2013		Notice to fend off Bitcoin risks.			Be the first nation to formally recognize bicoin [309].
2014	IRS issues financial legislation for New York.		"Mt.Gox" incident occurred [313].		UK strengthen to fend off illegal block-chain activity.
2015	NYDFS announces BitLicence for block-chain supervision.	People's bank of China investigates digital currency.			UK proposes the notion of regulatory sandbox [311].
2016	OCC announces responsible innovation framework [314].	Release blockchain white paper [315] and 13TH five year plan [310].	The Cabinet vote to agree on recognizing Bitcoin.		UK releases blockchain and beyond, white paper [316].
2017		USA congress establishes committee for blockchain decision making.	Promote payment Act and notice to fend off malicious ICO.		
2018	SEC announces to supervise digital asset and blockchain.	Notice to fend off financing risks from ICO and blockchain.	FSA constrains registration and ICO of blockchain.		Establish committee for encrypted digital assets.
2019	Congress holds hearings for Facebook and Libra [317].	Release the first Cryptographic legislation [318].	Pass multiple bills to strengthen blockchain supervision	Establish strategic Act for blockchain regulation	FCA releases guidance for encrypted digital assets.

9.3.2   Challenge & Open Problem for Regulation

The regulation of blockchain industry is of vital importance to its long term prosperity and sustainable development. During the years of global blockchain regulation, many problems have arisen regarding the efficiency and effectiveness [319]. Frankly speaking, a strictly enforced regulation will impair the blockchain industry, hurting the economy. This goes against the genuine intention of regulation. An adequately enforced blockchain regulation will encourage and stimulate the blockchain industry towards prosperity and economic success. There is no generic guidance for regulation, and it goes in line with local development and national strategy. However, there are many lessons learned from previous regulations works [319,320]. Based on relevant surveys and studies, we conclude that current regulatory works suffer from the shortage of technical solutions, intensive tasks, inflexible approaches, constant security breaches and lack of integral solution. We explain these open problems as follows:

1. **Lack of technical means:** Although many governments have frequently released relevant legislation for regulation, it cannot properly address attacks like the "DAO" [63] or "WannaCry" [307]. Superficially, regulation authorities (mostly financial sectors) do not possess technical skills to navigate blockchain affairs. When things happen, the report is their only selection. Inherently, there is a lack of either countermeasure or forecasting measure to cope with cryptocurrency [1] or other high-tech products. It calls for co-operation of cryptographic community with regulation authority to come up with a solution.

2. **The intensity of regulation tasks:** Previous regulation tasks are carried out by man-to-man report manner. This does not work so well for high-tech products because there are several blockchain products but few regulation staff. RegTech [321,322] is a notion of employing technology itself to supervise technology instead of men's power. This technique, feasible as it is, still relies on technical means to take effect (discuss in the first point as well).

3. **The inflexibility of regulation approach:** Regulation and development are short-term and long-term contradictions. In the short term, regulation inevitably decreases or prevent the development. However, it will surely bring prosperity to the entire industry in the long term. However, to simply fulfil the regulation tasks, local law enforcement authorities incline to use trivial solutions, banning all ICO projects [323], for instance. This is understandable when there is no technical means and too many tasks to accomplish. However, this will impair the blockchain industry in the long run.

4. **Constant security breaches:** With new computer viruses come and go, our cyber world is facing a long-term struggle with security attacks [324]. Hence, it is necessary to devise countermeasures from all angles

and periods. In other words, there should be a prediction, countermeasure, and remedy methods against potential attacks. These techniques should be applied to the blockchain product in a distinct period.

5. **Lack of integral solution:** Consider blockchain is a distributed public layer maintained by each user in the network; it is useless to alter or rewrite just one single copy at any user-side for regulatory purposes. In other words, like any regulation action, all law enforcement authorities should work together to manage blockchain, and the majority of users should agree on the new blockchain history [215]. Any unilateral effort will not achieve a regulation goal; there should be an integral operation to regulate blockchain services and products.

9.3.3 Regulatory SandBox and Deployment

To overcome some open problems in the blockchain and other FinTechs ([308]), the regulatory sandbox is proposed as a promising solution. The UK's regulatory sandbox is first proposed and later applied in many economically active areas, such as Singapore, Hong Kong, New York, etc. As a flexible regulation tool, regulatory sandbox offers loose regulation environment. So, it has received success from worldwide and adopted by many countries to support high-tech developments. We review the development of regulatory sandbox in several countries in Table 9.2. As shown in Table 9.2, all listed countries have actively involved in the experimenting and launching of national regulatory sandboxes to develop their local technologies and economies.

According to surveys [325], and studies [326] on the regulatory sandbox, we can summarize the open problems as follows.

Problems associated with the practical execution of regulatory sandbox are: (1) difficult to acquire the bank's participation and service. (2) suffer from low stability and efficiency of tests. (3) hard to guarantee the rights of customers. (4) require close participation with regulatory authorities and governments. (5) it is needed for tested companies to work in alliance with the bank during sandbox test [327].

Problems associated with the localization of regulatory sandbox are: (1) need the government and law enforcement authority to take the lead. (2) need to consider the status and development schedule of local market and industry. (3) need to work under the national plan and strategy for development.

Problems associated with the inherent limits of the regulatory sandbox are: (1) The truthfulness of current sandbox tests are doubtful. (2) Number of facilities to participate is limited. (3) Most test procedures are dependent on local laws [328].

TABLE 9.2: Global Development of Regulatory Sandbox

Year	UK	China	Singapore	Canada	USA
2015	Release RS and guidance [311].				
2016	Register the first group of candidates for RS test.	HKMA promotes RS for FinTech [329].	Release guidance for RS of Fintech.		Senator suggests for RS Act.
2017	Carry out RS tests.	Consider legitimate digital currency.	PolicyPal completes 6-months term of RS test [330].	Be the first nation to apply RS to ICO [331].	
2018	Suggest to establish global RS [332],	Hong kong join as the founder of GFIN.	Suggest to open fast track for application of RS.		State of Arizona implements local RS.
2019	GFIN [333] is officially established for international RS.	Release plan of FinTech development 2019-2021 for RS.			State of Arizona implements local RS.
2020		Publish the first list of RS candidates for enrolment			Hawaii State release the first crypto - currency -oriented RS.

RS: Regulatory Sandbox.

9.3.4 Redactable Blockchain

Mutable blockchain [215] is a novel idea to reverse transaction history and redact blockchains for good. Mutable blockchain (also known as redactable blockchain) implies the block or the transaction with an editable feature for security or privacy concerns. It seeks to bypass mutability or break mutability for justified reasons. To date, Reversecoin is the first cryptocurrency to support transaction revocation by using a configurable timeout and revocation keys. Until 2017, Ateniese et al. [215] formally proposed notion of redactable blockchain in EuroS&P 2017 as a countermeasure against the misuses and abuses of blockchain immutability (e.g. financing terrorists, facilitating black market trade, etc.). Another legitimate motivation is erasing individual privacy or unwanted information from the blockchain to fulfill "the right to be forgotten" demanded by General Data Protection Regulation (GDPR). Relevant works on mutability can be categorized as follows:

Chameleon Hash-based Work: This type of works focuses on the cryptographic study of the chameleon hash function. Ateniese et al. [215] formally proposed to use the chameleon hash as the core element to derive redactable blockchain. Ashritha et al. proposed splitting the trapdoor key (private key) among multiple validators and reconstructed it via multi-party computation (MPC) for secure redaction. As a result, redaction is safely performed among multiple entities without mutual trust.

Mutable Transaction or History: This type of work seeks for a non-cryptographic solution to derive redaction. Puddu et al. [284] proposed μchain to alter transaction state at the expenses of introducing high computing costs. Deuber et al. proposed an efficient and redactable blockchain by consensus-based voting and policy to dictate the redaction. Only tiny overhead is yielded for their proposal in comparison with other works.

To summarize, the chameleon hash is a hot topic for the study of mutability. The chameleon hash is the basic building block of many other primitives, such as commitment, deniable signature, online and off-line signature, etc. Since digital signature follows hash-and-sign paradigm, a chameleon hash can be integrated with the signature scheme naturally without altering or impairing blockchain's original design. Therefore, our proposal in this work is plausible and compatible with the blockchain.

9.4 Economics

Works on blockchains from economic perspective fall into following categories: Pricing [334–340]; ICO [341,342]; Ponzi [343,344]; Study of Ethereum's economy [345–351]; Business process [352–360]; Economic application and system [361–364]. Here, ICO means Initial Coin Offering of cryptocurrencies. Ponzi is a typical business fraud.

Generally, economies of blockchains closely relate to the economic activities associated with Bitcoin, Ethereum and other cryptocurrencies. These activities include pricing, trade, investment, etc, of blockchains. To note, ICO is the starting point of the cryptocurrency project, which often attracts massive investment. Among these investments, Ponzi scandals (business fraud) may occur to disrupt market order.

9.4.1   Pricing

Pricing of the blockchain (e.g. Ethereum) is of vital importance to its users and investors. Following works [334–340] study pricing issues which affect Ethereum; ref. [337] studies the prediction of Ether price. We give an overview of pricing-based works in Table 9.3.

TABLE 9.3: Works for Pricing

Reference	Topic	Problem	Solution
Sovbetov et al. [334]	Coin price	Interpreting pricing factors	ARDL technique and several findings
Corbet. [335]	Coin price	Pricing bubbles	Find evidence to prove Bitcoin is in bubble phase
Mensi et al. [336]	Coin price	Price Fluctuation	Adopt Bai and Perron's structural change model
Poongodi et al. [337]	Coin price	Price Prediction	Perform price prediction with machine learning
Bouri et al. [338]	Coin Price	Price explosivity	Show a multi directional co-explosivity behavior
Sy and Morris [339]	Coin Price	Price fluctuation	Give pricing models and adopt learning method

Sovbetov et al. [334] examined factors that influence prices of most common five cryptocurrencies such as Bitcoin, Ethereum, Dash, Litecoin, and Monero over 2010–2018 using weekly data. The study employed the ARDL technique and documented several findings. They showed that cryptomarket-related factors such as market beta, trading volume, and volatility appear to be a significant determinant for all five cryptocurrencies. Meanwhile, the attractiveness of cryptocurrencies also matters in terms of their price determination, but only in long-run. What is more, error-correction models for Bitcoin, Ethereum, Dash, Litcoin, and Monero show that cointegrated series cannot drift too far apart.

Corbet [335] provided insight into the relationship between the relationship of cryptocurrency pricing discovery and internal fundamental explanatory variables that can generate the conditions and environment in which a pricing bubble can thrive. Corbet claimed to find evidence that supports the view that

Bitcoin is currently in a bubble phase and has been since the price increased above 1,000 US dollars.

Mensi et al. [336] explored the impacts of structural breaks (SB) on the dual long memory levels of Bitcoin and Ethereum price returns. This work is motivated by the large, frequent price fluctuation and excessive volatility observed in the cryptocurrency market, Mensi et al. adopted Bai and Perron (2003)'s structural change model. The results may indicate to financial practitioners or researchers to realize the possible instability of parameters in all aspects of cryptocurrency analysis and modeling due to the frequent existence of change points affected by underlying internal or external factors.

Poongodi et al. [337] performed price prediction with two machine learning methods, namely linear regression (LR) and support vector machine (SVM), by using a time series consisting of daily ether cryptocurrency closing prices. The results indicated that their proposed SVM model without additional features had a 10.62% higher accuracy (96.06%) than the LR method (85.46%). The accuracy difference further increased to 13.54% when adding additional features into the SVM algorithm. Even during a price rise, the proposed models would fail only once when the price suddenly rises. After that, the last three values would reflect a rise in price and the fourth value can be easily inferred. This accuracy is extremely suitable because only the next mean value must be predicted.

Bouri et al. [338] date stamped the price explosivity in leading cryptocurrencies and revealed that multiple explosivities characterised all cryptocurrencies investigated herein. Concretely, Bouri et al. studied price explosivity in leading cryptocurrencies. They found explosivity in one cryptocurrency leads to explosivity in another. Results showed evidence of a multidirectional co-explosivity behavior that is not necessarily from bigger to smaller and younger markets.

Sy and Morris [339] provided the development of a promising initial prototype pricing model for Bitcoin, Ethereum, and Litecoin. The proposed pricing models resulted in an average 7% difference between actual and predicted price for Bitcoin and Ethereum, and a 4% difference for Litecoin along a timeline, through the use of machine learning and deep learning, artificial neural networks using the contributing factors of the key variables and how they influence and capture pricing and investor behavior. They also identified additional datasets, such as sentiment market data into the model, along with larger exploration of blockchain and raw transaction mining, to increase the model's accuracy and forecasting ability.

Liu et al. [340] presented a blockchain-enhanced data market framework and an optimal pricing mechanism for IoT. They proposed an edge/cloud-computing-assisted, blockchain-enhanced IoT data market framework. They designed an optimal pricing mechanism to support efficient trading in the IoT data market using a game theory-based model. They formulated a pricing and purchasing problem between the data consumer and the market-agency that sets prices for data owners. They adopted a two-stage Stackelberg game to

TABLE 9.4: Study of Ethereum's Economy

Ref	Topic	Problem	Solution
Davidson et al. [345]	Conceptual Review	Governance of blockchain economics	Case study of Ethereum-based infrastructure
Davidson et al. [346]	Conceptual review	Institutional governance	Case study of Backfeed
Swan [349]	Conceptual review	Integration of artificial intelligence	Use deep learning concept
Scott [350]	Conceptual review	Technologies' potential	A primer on Bitcoin basics with suggestions
Bohme et al. [347]	Conceptual review	Design principles	Present the platforms design principles
Tikhomirov [348]	Technical review	Open challenges	Give technical overview of Ethereum
Norman [351]	Investment of ether	Investment risks	Discuss how to invest money in Ethereum

maximize the profits of the data consumer and the market-agency and proposed a CPS that considers the competition between data owners. Through backward induction, they investigated the consumer's optimal purchasing strategy in the second stage and the pricing strategies of the market-agency in the first stage. They proved that the Stackelberg equilibrium is derived analytically. We conducted numerical simulations to evaluate the proposed pricing scheme's performance, which shows the effectiveness and efficiency of the CPS compared with the IPS. With competition-enhanced pricing, the consumer can purchase more data at a lower price, and the win-owner can also efficiently gain more economic profit.

9.4.2   Research of ICO in Blockchains

Liu et al. [365] examined 1388 ICOs, published on December 31, 2017, on icobench.com website. They defined success criteria for the ICOs, based on the funds gathered, and on the behavior of the related tokens' price, finding the factors that most likely influence the ICO success likeliness.

The analysis showed that some factors are correlated to an ICO success. It looks that ICOs come from Slovenia and USA, and, to a lesser extent, Israel and China, are more prone to success. Most other countries do not bear the significance. The team size does not seem to be relevant to success. A high overall rating on icobench.com site, on the other hand, looks quite correlated to the success of an ICO, though this looks mainly due to the robot's advice rather than to the human experts' advice. Finally, managing the ICO token on Ethereum blockchain looks another success factor.

Fenu et al. [366] assessed the determinants of the amount raised in 423 ICOs. They discussed "token economics and limitations in ICO structuring, giving rise to conflicts of interest and exposing investors to significant risks. They analysed issues around valuation, accounting and allocation of value, as well as trading of tokens. They highlighted the importance of network effects as a source of value creation in ICO offerings and the limits to the use of ICOs as a financing tool.

9.4.3 Ponzi Scheme in Blockchains

Ponzi scheme is notorious economic scandals, and illegal activities in blockchains [343,344].

TABLE 9.5: Study of Business Process Execution

Reference	Topic	Problem	Solution
Lopez-Pintado et al. [352]	Technical study	Combining techniques in platform	Introduce a blockchain-based BPMN execution engine
Lopez Pintado et al. [353]	Technical study	Process tracking	Propose an open-source system: Caterpillar
Haarmann et al. [354]	Technical study	Decision taking	Evaluate by proof-of-concept prototype
Rimba et al. [355]	Technical study	Blockchain cost	Investigate the cost of chain by business process execution
Meironke et al. [356]	Technical study	Compliance challenges	Categorize business process compliance challenges
Banuelos et al. [357]	Technical study	Process optimization	Propose an optimized method for business processes
Kapitonov [358]	Technical study et al. [358]	Communication organization	Propose method for autonomous business activity
Di Ciccio et al. [359]	Technical study	Context of supply chain	Propose to provide full traceability of run-time enactment
Weber et al. [360]	Technical study	Trust in process execution	Run experiments for demonstration of feasibility

Chen et al. [343] proposed an approach to detect Ponzi schemes on the blockchain using data mining and machine learning methods to prevent attracting scams. By verifying smart contracts on Ethereum, they first extracted features from user accounts and operation codes of the smart contracts and then build a classification model to detect latent Ponzi schemes implemented as smart contracts. Using the proposed approach, they estimated more than 400 Ponzi schemes are running on Ethereum. Based on these results, they proposed to build a uniform platform to evaluate and monitor every created smart contract for early warning of scams.

Massacci et al. [344] showed that the failure of security property, e.g. anonymity, can destroy a DAOs because economic attacks can be tailgated to security attacks. Based on the proposed approach, they estimated that more than 400 Ponzi schemes are running on Ethereum. Based on these results, they proposed to build a uniform platform to evaluate and monitor every created smart contract for early warning of scams.

9.4.4   Study of Blockchain Economies

Following works study economies of blockchain in [345–351].

For a conceptual review, Davidson et al. [345] suggested two approaches to study the economics of blockchain: innovation-centred and governance-centred. They argued that the governance approach-based in new institutional economics and public choice economics is promising because it models blockchain as a new technology for creating spontaneous organizations, i.e. new types of economies. They illustrate this with a case study of the Ethereum-based infrastructure protocol and platform Backfeed.

Davidson et al. [346] presented the view of blockchains as new institutional technology of governance through a case study of Backfeed, an Ethereum-based platform for creating new types of commons-based collaborative economies.

Swan [349] discussed blockchain distributed ledgers in the context of public and private Blockchains, enterprise blockchain deployments, and the role of blockchains in next-generation artificial intelligence systems, notably deep learning blockchains.

Scott [350] provided a primer on the basics of Bitcoin and the existent narratives about the technology's potential to facilitate remittances, financial inclusion, cooperative structures and even micro-insurance systems. He concluded this work with suggestions for future research.

Bohme et al. [347] presented the platform's design principles and properties of Ethereum for a non-technical audience; reviews its past, present, and future uses and points out risks and regulatory issues as Bitcoin interacts with the conventional financial system and the real economy.

For technical issues of Ethereum, Tikhomirov [348] summarized the state of knowledge in Ethereum. He provided a technical overview of Ethereum, outlined open challenges, and reviewed proposed solutions. He also mentioned alternative smart contract blockchains.

To review investment issues of Ethereum's Economy, relevant research is [351]. This book will help the author learn the following: Which way of making money in the cryptocurrency market suits the author best. Where should the reader start if the reader has just 500 US dollars.

9.4.5 Business Process Execution

Business process execution [352–360] is a research topic closely related to blockchain economies.

Lopez-Pintado et al. [352] demonstrated how to combine the advantages of a business process management system with those of a blockchain platform. They introduced a blockchain-based BPMN execution engine, named Caterpillar. Like any BPMN execution engine, Caterpillar supports creating instances of a process model and allows users to monitor the state of process instances and to execute tasks thereof. The Caterpillar compiler supports many BPMN constructs, including sub-processes, multi-instance activities and event handlers. They described Caterpillar's architecture and the interfaces it provides to support the monitoring of process instances, the allocation and execution of work items, and the execution of service tasks.

In another work, Lopez-Pintado et al. [353] proposed a demonstration to introduce Caterpillar, an open-source Business Process Management System (BPMS) that runs on top of the Ethereum blockchain. Like any BPMS, Caterpillar supports creating instances of a process model and allows users to track the state of process instances and to execute tasks thereof.

Haarmann et al. [354] argued that decisions are an essential aspect of interacting business processes, and, therefore, need to be executed on the blockchain. The interacting processes can use the immutable representation of decision logic to be more secure, transparent, and better auditable. The approach is based on mapping the DMN language S-FEEL to Solidity code to be run on the Ethereum blockchain. The work is evaluated by a proof-of-concept prototype and an empirical cost evaluation.

Rimba et al. [355] investigated the cost of blockchain using business process execution as a lens. Specifically, they compared the cost for computation and storage of business process execution on blockchain vs a popular cloud service. They captured the cost models for both alternatives. Then, they implemented and measured the cost of business process execution on blockchain and cloud services for an example business process model from the literature. They observed two orders of magnitude difference in the cost.

Meironke et al. [356] identified 21 business process compliance (BPC) challenges and categorized these into legal, organizational, human-centred, technical and economic challenges. They found that the technical and organizational BPC challenges were those that Ethereum could best solve, while human-centred challenges could be less well addressed. Furthermore, the Ethereum blockchain's implementation leads to additional challenges, such as the immutability of illegal content within the Ethereum blockchain or the error-proneness and zero-defect tolerance of smart contracts.

Garcia-Banuelos et al. [357] proposed an optimized method for executing business processes on top of commodity blockchain technology. Their optimization targets three areas specifically: initialization cost for process instances, task execution cost using a space-optimized data structure, and improved runtime components for maximized throughput. The method is empirically compared to a previously proposed baseline by relaying execution logs and measuring resource consumption and throughput.

Kapitonov et al. [358] showed how to organize a communication system between agents in a peer-to-peer network using the decentralized Ethereum Blockchain technology and smart contracts. The architecture of the protocol of autonomous business activity, based on this communication method, is given. As a result, implementing an autonomous economic system with unmanned aerial vehicles (UAV) is described.

Di Ciccio et al. [359] investigated how to run a business process in the context of a supply chain on a blockchain infrastructure to provide full traceability of its runtime enactment. Their approach retrieves the information to trace the process instances' execution solely from the transactions written on-chain. To do so, hash-codes are reverse-engineered based on the Solidity Smart Contract encoding of the generating process. They showed the investigation results through an implemented software prototype, with a case study on the reportedly challenging context of the pharmaceutical supply chain.

Weber et al. [360] addressed the fundamental problem of trust in collaborative process execution using blockchain. They ran a large number of experiments to demonstrate this approach feasibility, using a private and public blockchain. While latency is low on a private, customized blockchain, the latency on the public blockchain may be considered too high for fast-paced scenarios. Additional benefits of their approach include the option to build escrow and automated payments into the process, and that the blockchain transactions from process executions form an immutable audit trail.

9.4.6   Application associated with Blockchain Economies

Applications built for economic activity in Ethereum platform are studied in Ref. [361–364].

Hai Luong. [361] studied how the blockchain network can be integrated into a social, financial mobile application. The research is completed by developing a smart contract and connecting it with the mobile application. The smart contract is written in the Solidity programing language and run on the Ethereum network.

Biryukov et al. [362] introduced Findel, a purely declarative financial domain-specific language (DSL) well suited for implementation in blockchain networks. They introduced Findel a declarative financial DSL built upon ideas from previous research in financial engineering. Formalizing contract clauses using Findel makes them unambiguous and machine-readable. They proved Ethereum to be a suitable platform for trading and executing Findel contracts. Nevertheless, the whole smart contract is still in its infancy. Programers who wish to implement a usable smart contract for handling financial agreements need to be aware of the future challenges: from fundamental limitations of the blockchain network architecture to imperfect development environment. After modeling and analyzing existing business processes.

Kopylash [363] created a proof-of-concept of a hybrid system that integrates the Ethereum smart contract and traditional web application. Also, Kopylash introduced the concept of tampering-resilient document storage and extended the baseline solution to add support for such storage-based on IPFS. Finally, Kopylash summarized and discussed the potential problems that can be met during the development of a blockchain-based application. Kopylash provided potential solutions and described their implications. To sum up, this paper's result is the implementation of a proof-of-concept of a hybrid application that integrates Ethereum smart contracts with a traditional web application. The implementation is based on a case study of a Singaporean real estate company and is done in a tight collaboration with the company (from the business side). This work summarized the experience of applying technology to a case study and gives the overview of potential problems and possible solutions during the development of a blockchain-based real estate application.

Beck et al. [364] assembled the first batch of IS research papers on blockchain technology. The special issue assembled the first batch of IS research papers on blockchain technology. They believed that the research portfolio provides a springboard for aspiring IS researchers to start new research on the topic. The various examples of the use of blockchain in different industries demonstrate the broad applicability of the technology.

9.5 Chapter Summary

This chapter is devoted to investigating global development in regulation, and flexible and practical countermeasures to achieve blockchain regulation goals. It categorizes and reviews business activities closely related to blockchain. This is useful for the researcher to learn global developments in blockchain regulation. It also helps the practitioner to learn the business patterns and potential weak links in conducting blockchain regulation affairs. After reading this chapter, the reader should: (1) Know global developments in blockchain regulation. Know what is the regulatory sandbox. (2) Know challenges and possible solutions for blockchain regulation. (3) Know business events and affairs associated with blockchain services. To summarize, current regulatory works suffer from the shortage of technical solutions, intensive tasks, inflexible approaches, constant security breaches and lack of integral solution. The global community is enforcing ever-increasingly flexible but accurate regulation on the blockchain industry. It calls for a unified regulatory solution to defend security attacks while providing real-time and efficient supervision on the blockchain. Our studies and surveys in this chapter help government and financial sectors improve their views and abilities in regulating the blockchain industry.

Chapter 10

Concluding Remarks

10.1 Summary of Design and Analysis

This book is mainly devoted to study the current designs and analysis of public key cryptographic algorithms for blockchain. Based on the introduction, preliminaries and background of blockchain and PKC, this book enumerates 14 cases to analyse design, security and efficiency. Among the 14 enumerated cases, four of them are standard, and textbook cases (i.e. ECDSA, BLS, IBE, CS), three of them are notable and milestone-like works (i.e. GS08, GR16, CA17), six of them are the book authors' original works, two of them are other notable works.

We summarize all enumerated works in Table 10.1 regarding their designs, security and proof techniques. As observed in Table 10.1, we can see that it is not trivial to build security reduction for a given scheme. The proof technique is closely related to specific intractability based on which the scheme is built and researchers' habit to programe the relevant security reduction. The former implies a highly logical relationship between security analysis and specific mathematical intractable problem. The latter indicates strong empirical knowledge in analyzing the security of a given cryptographic scheme. Up to date, we are not able to give further investigation or assert the exact use of the proof technique for a given scheme under specific intractable problem, but we do argue that this knowledge and skills can be learned and imitated in a right way through repeated practices and clear understanding. Next, we summarize each category of enumerated schemes.

1. **PKS.** Signing is the most preliminary and representative cryptographic primitive used in the current blockchains. Early designs, being short in signature (e.g. ECDSA) or simple in construction (e.g. BLS), generally contribute to blockchain's early success. However, the versatility of blockchain and citizens' awareness of privacy have driven industries in a quest of secure group signature and ring signature. Compared with ECDSA or BLS, the group and the ring signatures are more complex in design and generally yield a linear signature size with the chosen ring. Meanwhile, the proofs of these schemes are sophisticated and will be even more if considering threshold (e.g. multiple signers) or protocol (running interactively) scenarios. In other words, to gain empirical

TABLE 10.1: Summaries of designs and security

Scheme	Design	Security &Proof Techniques
Public-Key Signature (PKS) for Blockchain		
ECDSA [100]	Short signature size and DLP-based	EU-CMA secure & Proof by contradiction.
BLS [31]	Simple and CDHP-based	EU-CMA secure & Proof by sequences of games.
MLSAG [102]	Linear signature size with chosen ring	EU-CMA secure & Rewind simulation [130].
LRRS [132]	Linear signature size with chosen ring	EU-CMA secure & Rewind simulation, proof by redaction
Public-Key Encryption (PKE) for Blockchain		
IBE [156]	Based on Hybrid encryption [168]	IND-CCA2 secure in ROM & Proof by transitions of failure events.
CS [169]	Practical design based on DDHP	IND-CCA2 secure in standard model & Proof by contradiction.
HDRS [190]	Support signcryption but is inefficient in Decryption	IND-CPA secure in ROM & Proof by transitions of indistinguishability.
Public-Key Hash (PKH) for Blockchain		
ACH [221]	Pairing-based and CDHP-based	EU-CA secure, IND-FA secure & Transitions based on failure event.
HCCH [190]	Homomorphic and DDHP-based	EU-CA secure, IND-FA secure & Proof by transitions of indistinguishability.
PCH	Pairing-based and DDHP-based	IND-D&P secure, Col-UNP secure & Proof by transitions of indistinguishability.
TUCH [132]	Pairing-based and CDHP-based	EU-CA secure, IND-FA secure & Rewind simulation [130].
RCH [224]	Pairing-based and CDHP-based	EU-CA secure, IND-FA secure & Proof by transitions based on indistinguishability.
Zero-Knowledge Proof (ZKP) for Blockchain		
GS08 [225]	Bilinear groups-based	NIWI & Proof by illation.
GR16 [226]	Pairing-based NIZK	Disclosure-freeness & Case analysis.
CA17 [227]	Revealing attacks from modifying CRS	Strengthen by checking CRS & n/a.

EU-CMA: existentially unforgeable against a chosen message attack; n/a: not applicable; DLP: discrete logarithm problem; IND-CCA2: indistinguishability against adaptively chosen message attack; ROM: random oracle model; EU-CA: as existentially unforgeable against collision attack; IND-FA: indistinguishability of forgery attack; IND-D& P: indistinguishability between deterministic and probabilistic hashes; Col-UNP: collision-resistance under unpredictable file source; NIWI: non-interactive witness indistinguishability; CRS: common reference string.

knowledge of using a proof technique, the reader is referred to study more cases of classical design and rigorous proof.

2. **PKE.** Encryption is the basic means and primitive to protect data confidentiality. Although most PKEs are not efficient enough for real-time application use, it is widely suggested to achieve IND-CCA2 security for any practical use. Although our HDRS scheme is not IND-CCA2 secure, it instantiates how the proof technique is specifically practiced. The instantiated IBE and CS schemes in this book train, readers, to be an insightful designer for encryption. Following the generic methods, in Section 5.7, readers can build IND-CCA2 schemes in various ways. By familiarity with the proof technique and the generic construction of IND-CCA2, the reader can always start from the basic building block to a full and practical encryption scheme while proving its security in a rigorous way. Though we cannot conclude an overall representative PKE scheme for use in blockchain, we assert that any IND-CCA2 schemes with rigorous proof are eligible candidates for blockchain from a security perspective.

3. **PKH.** Hashing is the fundamental tool for blockchain; it sequentially links each block forming a chain. Unlike symmetric hash function, we formalize a special type of hash scheme, called public key hash (also known as: chameleon hash, CH). Based on our previous efforts in PKH, we instantiate multiple designs of CH and some ways to practice proof techniques. We show our empirical knowledge in designing and analyzing CH based on fruitful case analysis and personal experience. The security of CH exploits some security like indistinguishability-type of encryption and unforgeability-type of signature. So, the lecture from PKH given to the reader helps the reader grasp and master all proof techniques taught in previous studies. Moreover, PKH is a trapdoor one-way hash function; it serves as building blocks for NIZK, IND-CCA2 security, etc. Therefore, the studies of PKH is an important topic in the design and analysis of PKC and blockchain.

4. **ZKP.** Zero-Knowledge proof is a typical primitive to study zero information leakage during a proof. It has been used in ZCash and increasingly involved in the design of modern cryptocurrencies. We show generic ways to construct zk-SNARK. Specifically, although zk-SNARK is generally considered as a practical tool for blockchain, it is not trivial to design. In fact, the construction process is highly sophisticated and impractical to deploy in real application. From the analysis of three cases, we conclude that security challenges still exist to introduce zk-SNARK to blockchain even if we bypass the complex building procedure. For example, the generation of CRS suffers from malicious attacks, and an honest generator's assumption cannot be erased easily. Meanwhile, it is difficult for the reader to step in the studies of ZKP without referring to historical

research or figuring out many technical terms like the algebraic circuit. This also coincides with our intention to put ZKP as our last chapter for case analysis: the reader should learn more knowledge from previous chapters to find out how ZKP is implemented and why in that way. To conclude, the studies of ZKP are more like an advanced and integrated learning stage based on previous studies. This process involves profound knowledge and is a long-term study for the reader to pursuit.

10.2   Open Problems

We summarize open problems as follows:

10.2.1   Challenges in Designing Practical PKS

In the design of practical public-key signature, signature size, security, signing and verification efficiency are all counted as essential factors. For signature size, a trivial problem is the linear signature size with the chosen group or ring. A linear size signature is unscalable and impractical to be deployed in the blockchain where the block size is fixed and small. Also, it calls for achieving more robust security. On the one hand, it asks the design to rely on as weaker assumption as possible (ideally, without random oracle assumption). On the other hand, it quests for security model to capture more practical attacks. For example, for our LRRS signature built from the chameleon hash, we should consider the security model which describes key exposure attack where adversary seeks to recover a signer's private key from two forged signatures. Also, regarding efficiency, as previously evaluated, signing generally costs less time than verification. However, due to the increased scalability of blockchain, it is recommended to decrease the verification costs dramatically. Therefore, the above imply a design of efficient, short and constant signature length with strong security for public key signature.

10.2.2   Challenges in Designing Practical PKE

It is generally recommended to achieve IND-CCA2 security for any PKE scheme to be deployed in the blockchain. However, according to our studies, most existing methods to construct IND-CCA2 security either require theoretical assumption (random oracle) or employ impractical techniques (e.g. NIZK) to achieve desired security. Also, the proof of a given IND-CCA2 encryption scheme is either highly logical to programe or relies on empirical knowledge to produce. This makes it difficult for a cryptographic beginner to complete a practical design and analysis. What is more, the design of PKE with practical security in the standard model is even more difficult to achieve, as it relies on sophisticated mathematical skills to design and carefully plotted analysis to validate. Another critical factor is the deficiency of overall PKE scheme. However, countermeasures like adopting symmetric encryption to encrypt data in the blockchain and utilizing asymmetric encryption to encrypt symmetric encryption keys exist. This results in further issues, e.g. key management, key escrow problems, etc. Therefore, devising a practically secure and efficient PKE is challenging.

10.2.3   Challenges in Designing Practical PKH

Based on several cases instantiated with our empirical knowledge, the reader should find it challenging to devise and analyze the PKH scheme. However, important security has not been yet formally analyzed: key exposure freeness. Although we managed to define it in our several works, we fail to give a strict and rigorous definition and build the design under specific intractable problem. We define it as a highly dependent issue with concrete algorithmic design and intractable assumption while leaving it as the future work. As some may define this notion by extraction from a pair of previously forged hash collision; others may capture this notion by using the trapdoor (secret information generated by the private key). In other words, it is non-trivial to define and solve the key-exposure problem. To add, since PKH is a trapdoor one-way hash function, each PKH naturally yields a trapdoor commitment hash function scheme. This helps us learn NIZK and other variants (e.g. zk-SNARK) since it adopts a commitment scheme to generate NIZK proof. To conclude, the studies of PKH is a good starting point for a beginner to start with. Based on our rich published works and instantiated cases, the reader is given a hieratical and constructive view of building PKH. Meanwhile, following these lectures of PKH, learners can quickly accumulate knowledge and develop skills to perform design and proof for many complex primitives such as encryption, ZKP, etc.

10.2.4   Challenges in Designing Practical ZKP

ZKP is a promising and popular primitive to build diverse and secure blockchains. However, a practical variant of ZKP is non-trivial work to fulfill. The studies of generic methods to construct zk-SNARK require patience and efforts from readers. Besides, according to our case analysis, ZKP schemes generally have highly technical terms and a history of continuous improvement for the relevant security. In other words, those who fail to follow the research line closely cannot easily interpret the inherent meaning or motivation of some minor security or technical designs. This makes a beginner full of questions when reading a new research work on ZKP. Meanwhile, although zk-SNARK receives much attention from researchers, there still exist problems in building a practical blockchain. One of the most concerning problems is the assumption of an honest setup stage to initiate parameters for ZKP in zk-SNARK. Although some works try to address this problem by removing the centralized trust for the initiation, it either brings high complexity to the overall system or introduces impractical assumption which weakens the scheme's security. To conclude, devising a practical zk-SNARK with totally decentralized initiation to generate parameters is still challenging.

10.2.5   Challenges in Optimizing Blockchain

The research on optimizing blockchain extends to security, privacy, efficiency, flexibility, etc. Recent years have witnessed some successful solutions like lightning network, micro-payment, etc. These works give satisfactory solutions to remove some bottlenecks of current blockchains at some point. Some techniques work as an overlay, and some are encapsulated in the inherent design of blockchains. However, it is not always successful in adopting these tools or techniques as solutions. For example, the involvement of multi-party computation can achieve secure interaction to some extents. However, it cannot deal with massive interactions when the number of participants in protocols surges. Another example is the use of the chameleon hash to achieve redactable blockchain. Although CH can transform blockchain to a mutable ledger to remove some malicious contents. It also brings additional security, privacy and efficiency concerns regarding how to implement the redaction in massive scale. To summarize, the smart and appropriate use of optimization tools determines the success of an optimization task. Any improper technique used as an overlay and mistakenly involved in the basic design will deteriorate and even generate new bottlenecks for the entire system. Therefore, how to implement an optimization tool to derive practical gains in challenging work.

10.2.6   Challenges in Blockchain Regulation & Blockchain Economies

The last decade has witnessed rapid responses from the global community to blockchain activities and businesses. Almost every developed country is paying close attention to and enforce strict regulation on the cryptocurrencies. According to our surveys, current regulatory jobs suffer from the shortage of technical solutions, intensive works, inflexible approaches, constant security attacks, and an integral solution. Among them, the shortage of technical solution and inflexibility are the most crucial problems. The regulatory sandbox is a leading mechanism proposed by the UK to perform flexible regulation. Although many countries have widely adopted it since 2015, it faces problems like localization of regulation; failures resulted from inherent designs, constant security attacks, etc. In other words, governments and law departments cannot trivially implement a regulatory sandbox mechanism to a local economy. Therefore, to summarize, successful implementation of regulation requires government sectors' participation, a design that fits local economies and without inherent flaws, etc. Achieving the above goals is challenging.

Also, blockchain is first and foremost utilized to serve financial services. However, due to decentralization and anonymity features, it is hard to trace and regulate the blockchain activities. Constant Ponzi incidents witnessed in ICO imply the chaos in implementing blockchain techniques in the financial field. How to supervise blockchain activities and lead them toward good ways are challenging works.

10.3 Conclusion

This book seeks to generalize the necessary techniques and experiences in designing and analyzing cryptographic schemes for blockchain. It devotes three chapters to review the background and basic knowledge, four chapters to discuss the specific type of cryptographic primitives designed for blockchain, one chapter to discuss optimization tool and another chapter for blockchain regulation and economies. This book covers a systematic survey of research objectives, detailed reviews of cryptographic schemes, lectures and methodologies to practice cryptography. We follow the hieratical level to introduce cryptographic knowledge, attempting to provide the learner with sufficient case analysis based on empirical knowledge. Specifically, this book enumerates 14 cryptographic schemes as use cases for analysis. Among the 14 covered cases, four of them are standard, and textbook cases (i.e. ECDSA, BLS, IBE, CS), three of them are notable and milestone-like works (i.e. GS08, GR16, CA17), six of them are the book authors' original works, and two of them are other notable works. Through step-by-step lecture and convenient pop-up tips as instructions, the reader is supposed to learn and practice some evaluation skills and proof techniques to design and analyse cryptographic schemes for blockchain. Based on our carefully plotted clues for analysis and arrangement of chapters from easy to difficult, the readers are supposed to learn public key signature, encryption, hashing, and zero knowledge systematically and intensively. The proof techniques instantiated in this book help the reader lay basic foundations to pursue studies and research in advanced cryptographic topics and fields.

This book summarizes the aforementioned schemes to derive some empirical knowledge and generalizes some open problems as future challenges. We conclude as follows: (1). The practical design and analysis of cryptographic schemes for blockchain can address significant problems in blockchain at the algorithmic level. This type of research has received popularity from both global cryptographic community and blockchain developer's community. This research field is the most fast-developing region of all research areas. (2). The intrinsic deficiencies in some traditional cryptographic primitives, like ring signature, IND-CCA2 secure encryption, zero-knowledge, etc, prevent the successful application of these primitives in the blockchain. However, tremendous efforts are being made to make these primitives practical and applicable by researchers. Hopefully, we can derive efficient and practically-secure zk-SNARKs or public key schemes as building blocks or overlays for blockchain in the near future. (3) The formal and rigorous design and analysis of public key cryptographic algorithms matters to the researches and development of blockchain. In cryptographic academia, it is generally recommended to design and analyze cryptographic schemes strictly by practicing provable security theory, complexity theory, and proof techniques to validate proposed schemes.

Bibliography

[1] Satoshi Nakamoto. Bitcoin: A Peer-to-Peer Electronic Cash System. Technical report, Manubot, 2019.

[2] Gavin Wood. Ethereum: A Secure Decentralised Generalised Transaction Ledger. *Ethereum Project Yellow Paper*, 151(2014):1–32, 2014.

[3] Whitfield Diffie and Martin Hellman. New Directions in *Cryptography*. *IEEE Transactions on Information Theory*, 22(6):644–654, 1976.

[4] Shafi Goldwasser, Silvio Micali, and Ronald L Rivest. A Digital Signature Scheme Secure against Adaptive Chosen-Message Attacks. *SIAM Journal on Computing*, 17(2):281–308, 1988.

[5] Mihir Bellare and Phillip Rogaway. Optimal Asymmetric Encryption. In *Workshop on the Theory and Application of Cryptographic Techniques*, pages 92–111. Springer, Berlin, Heidelberg, 1994.

[6] Charles Rackoff and Daniel R Simon. Non-Interactive Zero-Knowledge Proof of Knowledge and Chosen Ciphertext Attack. In *Annual International Cryptology Conference*, pages 433–444. Springer, Berlin, Heidelberg, 1991.

[7] John Malone-Lee. Identity-Based Signcryption. In *IACR Cryptology ePrint Archive*, volume 2002, 2002.

[8] Roberto Di Pietro and Alessandro Sorniotti. Boosting Efficiency and Security in Proof of Ownership for Deduplication. In *Proceedings of the 7th ACM Symposium on Information, Computer and Communications Security*, pages 81–82, 2012.

[9] Niels Ferguson and Bruce Schneier. *Practical Cryptography*, volume 141. Wiley, New York, 2003.

[10] Ian Miers, Christina Garman, Matthew Green, and Aviel D Rubin. Zerocoin: Anonymous Distributed E-Cash from Bitcoin. In *2013 IEEE Symposium on Security and Privacy*, pages 397–411. IEEE, Berkeley, CA, USA, 2013.

[11] Shen Noether. Ring Signature Confidential Transactions for Monero. *IACR Cryptology ePrint Archive*, 2015:1098, 2015.

[12] Shi-Feng Sun, Man Ho Au, Joseph K Liu, and Tsz Hon Yuen. Ringct 2.0: A Compact Accumulator-based (Linkable Ring Signature) Protocol for Blockchain Cryptocurrency Monero. In *European Symposium on Research in Computer Security*, pages 456–474. Springer, Cham, 2017.

[13] Mihir Bellare. Practice-Oriented Provable-Security. In *International Workshop on Information Security*, pages 221–231. Springer, Berlin, Heidelberg, 1997.

[14] Peter Bürgisser, Michael Clausen, and Mohammad A Shokrollahi. *Algebraic Complexity Theory*, volume 315. Springer Science & Business Media, 2013.

[15] Douglas Robert Stinson and Maura Paterson. *Cryptography: Theory and Practice*. CRC Press, 2018.

[16] Alfred J Menezes, Jonathan Katz, Paul C Van Oorschot, and Scott A Vanstone. *Handbook of Applied Cryptography*. CRC Press, 1996.

[17] Jonathan Katz and Yehuda Lindell. *Introduction to Modern Cryptography*. CRC Press, 2014.

[18] Juan Garay, Aggelos Kiayias, and Nikos Leonardos. The Bitcoin Backbone Protocol: Analysis and Applications. In *Annual International Conference on the Theory and Applications of Cryptographic Techniques*, pages 281–310. Springer, Berlin, Heidelberg, 2015.

[19] Philip Anderson. Perspective: Complexity Theory and Organization Science. *Organization Science*, 10(3):216–232, 1999.

[20] Steven M Manson. Simplifying Complexity: A Review of Complexity Theory. *Geoforum*, 32(3):405–414, 2001.

[21] Fuchun Guo, Willy Susilo, and Yi Mu. *Introduction to Security Reduction*. Springer, 2018.

[22] Jean-Sébastien Coron. On The Exact Security of Full Domain Hash. In *Annual International Cryptology Conference*, pages 229–235. Springer, Berlin, Heidelberg, 2000.

[23] Alexander W Dent. A Note On Game-Hopping Proofs. *IACR Cryptol. ePrint Arch.*, 2006:260, 2006.

[24] Victor Shoup. Sequences of games: A Tool for Taming Complexity in Security Proofs. *IACR Cryptol. ePrint Arch.*, 2004:332, 2004.

[25] Christoph Sprenger, D Basin, M Backes, Birgit Pfitzmann, and Michael Waidner. Cryptographically Sound Theorem Proving. In *19th IEEE Computer Security Foundations Workshop (CSFW'06)*, 14 pp. IEEE, Venice, Italy, 2006.

[26] Mihir Bellare and Phillip Rogaway. Random Oracles Are Practical: A Paradigm for Designing Efficient Protocols. In *Proceedings of The 1st ACM Conference on Computer and Communications Security*, pages 62–73, 1993. ACM, Fairfax, Virginia, USA.

[27] Craig Gentry and Zulfikar Ramzan. Eliminating Random Permutation Oracles in The Even-Mansour Cipher. In *International Conference on the Theory and Application of Cryptology and Information Security*, pages 32–47. Springer, Berlin, Heidelberg, 2004.

[28] Ran Canetti, Oded Goldreich, and Shai Halevi. The Random Oracle Methodology, Revisited. *Journal of the ACM (JACM)*, 51(4):557–594, 2004.

[29] Claus-Peter Schnorr. Efficient Signature Generation by Smart Cards. *Journal of Cryptology*, 4(3):161–174, 1991.

[30] David Chaum and Torben Pryds Pedersen. Wallet Databases with Observers. In *Annual International Cryptology Conference*, pages 89–105. Springer, Berlin, Heidelberg, 1992.

[31] Dan Boneh, Ben Lynn, and Hovav Shacham. Short Signatures from The Weil Pairing. In *International Conference on the Theory and Application of Cryptology and Information Security*, pages 514–532. Springer, Berlin, Heidelberg, 2001.

[32] Dan Boneh and Xavier Boyen. Short Signatures without Random Oracles. In *International Conference on the Theory and Applications of Cryptographic Techniques*, pages 56–73. Springer, Berlin, Heidelberg, 2004.

[33] Ali Dorri, Salil S Kanhere, Raja Jurdak, and Praveen Gauravaram. Blockchain for IoT Security and Privacy: The Case Study of A Smart Home. In *2017 IEEE International Conference on Pervasive Computing and Communications Workshops (PerCom Workshops)*, pages 618–623. IEEE, Kona, HI, USA, 2017.

[34] Zehui Xiong, Yang Zhang, Niyato Dusit, Ping Wang, and Han Zhu. When Mobile Blockchain Meets Edge Computing. *IEEE Communications Magazine*, 56(8):33–39, 2017.

[35] Charles Shen and Feniosky Pena-Mora. Blockchain for Cities - A Systematic Literature Review. *IEEE Access*, 6:76787–76819, 2018.

[36] Khaled Salah, M Habib Ur Rehman, Nishara Nizamuddin, and Ala Al-Fuqaha. Blockchain for AI: Review and Open Research Challenges. *IEEE Access*, 7:10127–10149, 2019.

[37] Mauro Conti, E. Sandeep Kumar, Chhagan Lal, and Sushmita Ruj. A Survey on Security and Privacy Issues of Bitcoin. *IEEE Communications Surveys & Tutorials*, 20(4):3416–3452, 2018.

[38] Amrit Kumar, Clment Fischer, Shruti Tople, and Prateek Saxena. A Traceability Analysis of Monero's Blockchain. In *European Symposium on Research in Computer Security*, pages 153–173. Springer, Cham, 2017.

[39] Zibin Zheng, Shaoan Xie, Hong Ning Dai, Xiangping Chen, and Huaimin Wang. Blockchain Challenges and Opportunities. *International Journal of Web and Grid Services*, 14(4):352–375, 2018.

[40] Huashan Chen, Marcus Pendleton, Laurent Njilla, and Shouhuai Xu. A Survey on Ethereum Systems Security: Vulnerabilities, Attacks and Defenses. *ACM Computing Surveys*, 53(3):1–43, 2020.

[41] Namecoin. Against Censorship. *https://www.namecoin.org/ (accessed Jan 14, 2021)*.

[42] Muneeb Ali, Jude Nelson, Ryan Shea, and Michael J Freedman. Blockstack: A Global Naming and Storage System Secured by Blockchains. In *2016 {USENIX} Annual Technical Conference ({USENIX}{ATC} 16)*, pages 181–194, 2016. USENIX, Austin, TX.

[43] Eleftherios Kokoris Kogias, Philipp Jovanovic, Nicolas Gailly, Ismail Khoffi, Linus Gasser, and Bryan Ford. Enhancing bitcoin security and performance with strong consistency via collective signing. In *25th {usenix} Security Symposium ({usenix} Security 16)*, pages 279–296, 2016. USENIX, Austin, TX.

[44] Monero. *https://www.getmonero.org/ (accessed Jan 14, 2021)*.

[45] Ian Miers, Christina Garman, Matthew Green, and A. D. Rubin. Zerocoin: Anonymous distributed e-cash from bitcoin. In *Security and Privacy (SP), 2013 IEEE Symposium on*, 2013. IEEE, San Francisco, California.

[46] Eli Ben Sasson, Alessandro Chiesa, Christina Garman, and et al. Zerocash: Decentralized Anonymous Payments from Bitcoin. In *2014 IEEE Symposium on Security and Privacy*, pages 459–474. IEEE, San Jose, CA, 2014.

[47] Bytecoin. The First Private Untraceable Crytocurrency. *https://bytecoin.org/ (accessed Jan 14, 2021)*.

[48] Counterparty. Counterparty Extends Bitcoin in New and Powerful Ways. *http://counterparty.io/ (accessed Jan 14, 2021)*.

[49] Dash. Your Money, Your Way. *http://dashpay.io/ (accessed Jan 14, 2021).*

[50] Litecoin. The Cryptocurrency for Payments. *https://litecoin.org/ (accessed Jan 14, 2021).*

[51] Facebooks Libra Currency to Launch Next Year in Limited Format. *https://www.ft.com/content/cfe4ca11-139a-4d4e-8a65-b3be3a0166be (accessed Jan 14, 2021).*

[52] Vitalik Buterin. Mastercoin: A Second-Generation Protocol On The Bitcoin Blockchain. *https://bitcoinmagazine.com/articles/mastercoin-a-second-generation-protocol-on-the-bitcoin-blockchain-1383603310 (accessed Jan 14, 2021).*

[53] Andrew Poelstra. Mimblewimble. *https://scalingbitcoin.org/papers/mimblewimble.pdf (accessed Jan 14, 2021).*

[54] Ripple. Move Money to All Corners of the World. *https://ripple.com/ (accessed Jan 14, 2021).*

[55] Shafi Goldwasser, Silvio Micali, and Charles Rackoff. The Knowledge Complexity of Interactive Proof Systems. *SIAM Journal on Computing*, 18(1):186–208, 1989.

[56] Austin Mohr. A Survey of Zero-Knowledge Proofs with Applications to Cryptography. *Southern Illinois University, Carbondale*, pages 1–12, 2007.

[57] Eng-Tuck Cheah and John Fry. Speculative Bubbles in Bitcoin Markets? An Empirical Investigation into The Fundamental Value of Bitcoin. *Economics Letters*, 130:32–36, 2015.

[58] Nicola Atzei, Massimo Bartoletti, and Tiziana Cimoli. A Survey of Attacks on Ethereum Smart Contracts (SoK). In *International Conference on Principles of Security and Trust*, pages 164–186. Springer, Berlin, Heidelberg, 2017.

[59] John R Douceur. The Sybil Attack. In *International Workshop on Peer-to-Peer Systems*, pages 251–260. Springer, Berlin, Heidelberg, 2002.

[60] Ethan Heilman, Alison Kendler, Aviv Zohar, and Sharon Goldberg. Eclipse Attacks on Bitcoins Peer-to-Peer Network. In *24th {USENIX} Security Symposium ({USENIX} Security 15)*, pages 129–144, 2015. USENIX, Washington, D. C.

[61] Samiran Bag, Sushmita Ruj, and Kouichi Sakurai. Bitcoin Block Withholding Attack: Analysis and Mitigation. *IEEE Transactions on Information Forensics and Security*, 12(8):1967–1978, 2016.

[62] Ting Chen, Xiaoqi Li, Ying Wang, and et al. An Adaptive Gas Cost Mechanism for Ethereum to Defend against Under-Priced DoS Attacks. In *International Conference on Information Security Practice and Experience*, pages 3–24. Springer, Cham, 2017.

[63] Muhammad Izhar Mehar, Charles Louis Shier, Alana Giambattista, and et al. Understanding A Revolutionary and Flawed Grand Experiment in Blockchain: The DAO Attack. *Journal of Cases on Information Technology (JCIT)*, 21(1):19–32, 2019.

[64] Tim Güneysu and Amir Moradi. Generic Side-Channel Countermeasures for Reconfigurable Devices. In *International Workshop on Cryptographic Hardware and Embedded Systems*, pages 33–48. Springer, Berlin, Heidelberg, 2011.

[65] John Gantz and David Reinsel. The Digital Universe in 2020: Big Data, Bigger Digital Shadows, and Biggest Growth in The Far East. *IDC iView: IDC Analyze the Future*, 2007(2012):1–16, 2012.

[66] Michael Backes and Dennis Hofheinz. How to Break and Repair A Universally Composable Signature Functionality. In *International Conference on Information Security*, pages 61–72. Springer, Berlin, Heidelberg, 2004.

[67] Giuseppe Ateniese and Breno de Medeiros. On The Key Exposure Problem in Chameleon Hashes. In *International Conference on Security in Communication Networks*, pages 165–179. Springer, Berlin, Heidelberg, 2004.

[68] Xiaofeng Chen, Fangguo Zhang, Willy Susilo, and Yi Mu. Efficient Generic On-Line/Off-Line Signatures without Key Exposure. In *International Conference on Applied Cryptography and Network Security*, pages 18–30. Springer, Berlin, Heidelberg, 2007.

[69] Mihir Bellare and Adriana Palacio. Protecting against Key-Exposure: Strongly Key-Insulated Encryption with Optimal Threshold. *Applicable Algebra in Engineering, Communication and Computing*, 16(6):379–396, 2006.

[70] Ahmed Kosba, Andrew Miller, Elaine Shi, Zikai Wen, and Charalampos Papamanthou. Hawk: The Blockchain Model of Cryptography and Privacy-Preserving Smart Contracts. In *2016 IEEE Symposium on Security and Privacy (SP)*, pages 839–858. IEEE, San Jose, California, 2016.

[71] A secure cryptocurrency scheme based on post-quantum blockchain.

[72] Divesh Aggarwal and Ueli Maurer. Breaking RSA Generically Is Equivalent to Factoring. In *Annual International Conference on the Theory and Applications of Cryptographic Techniques*, pages 36–53. Springer, Berlin, Heidelberg, 2009.

[73] Xiaoyun Wang and Hongbo Yu. How to Break MD5 and Other Hash Functions. In *Annual International Conference on The Theory and Applications of Cryptographic Techniques*, pages 19–35. Springer, Berlin, Heidelberg, 2005.

[74] Ghassan O Karame, Elli Androulaki, and Srdjan Capkun. Double-Spending Fast Payments in Bitcoin. In *Proceedings of The 2012 ACM Conference on Computer and Communications Security*, pages 906–917, 2012. ACM, Raleigh, NC, USA.

[75] David Chaum and Eugène Van Heyst. Group Signatures. In *Workshop on the Theory and Application of of Cryptographic Techniques*, pages 257–265. Springer, Berlin, Heidelberg, 1991.

[76] Ronald L Rivest, Adi Shamir, and Yael Tauman. How to Leak A Secret. In *International Conference on the Theory and Application of Cryptology and Information Security*, pages 552–565. Springer, Berlin, Heidelberg, 2001.

[77] Patrick P Tsang and Victor K Wei. Short Linkable Ring Signatures for E-Voting, E-Cash and Attestation. In *International Conference on Information Security Practice and Experience*, pages 48–60. Springer, Singapore, 2005.

[78] Andrew C Yao. Protocols for Secure Computations. In *23rd Annual Symposium on Foundations of Computer Science (SFCS 1982)*, pages 160–164. ACM, NW Washington, DC, United States, 1982.

[79] Oded Goldreich. Secure Multi-Party Computation. *Manuscript. Preliminary version*, 78, 1998.

[80] Yannan Li, Guomin Yang, Willy Susilo, Yong Yu, Man Ho Au, and Dongxi Liu. Traceable Monero: Anonymous Cryptocurrency with Enhanced Accountability. *IEEE Transactions on Dependable and Secure Computing*, 2019. volume 18, issue 2, pp 679 - 691.

[81] Guy Zyskind, Oz Nathan, and Alex Pentland. Enigma: Decentralized Computation Platform with Guaranteed Privacy. *arXiv preprint arXiv:1506.03471*, 2015.

[82] Ronald L Rivest, Len Adleman, Michael L Dertouzos, et al. On Data Banks and Privacy Homomorphisms. *Foundations of secure computation*, 4(11):169–180, 1978.

[83] Craig Gentry and Dan Boneh. *A Fully Homomorphic Encryption Scheme*, volume 20. Stanford University, 2009.

[84] Nicolas Gisin, Grégoire Ribordy, Wolfgang Tittel, and Hugo Zbinden. Quantum cryptography. *Reviews of Modern Physics*, 74(1):145, 2002.

[85] Hugo Krawczyk and Tal Rabin. Chameleon Hashing and Signatures. 1998.

[86] Jan Camenisch, David Derler, Stephan Krenn, Henrich C Pöhls, Kai Samelin, and Daniel Slamanig. Chameleon-Hashes with Ephemeral Trapdoors. In *IACR International Workshop on Public Key Cryptography*, pages 152–182. Springer, Amsterdam, Netherlands, 2017.

[87] Reversecoin. Worlds First Crypto-currency with "Reversible Transactions". *http://www.reversecoin.org/ (accessed Jan 14, 2021)*.

[88] George Bissias, A Pinar Ozisik, Brian N Levine, and Marc Liberatore. Sybil-Resistant Mixing for Bitcoin. In *Proceedings of the 13th Workshop on Privacy in the Electronic Society*, pages 149–158, 2014.

[89] Ren Zhang and Bart Preneel. Publish or Perish: A Backward-Compatible Defense against Selfish Mining in Bitcoin. In *Cryptographers Track at the RSA Conference*, pages 277–292. Springer, San Francisco, 2017.

[90] Ethan Heilman. One Weird Trick to Stop Selfish Miners: Fresh Bitcoins, A Solution for The Honest Miner. In *International Conference on Financial Cryptography and Data Security*, pages 161–162. Springer, Barbados, 2014.

[91] Iddo Bentov, Charles Lee, Alex Mizrahi, and Meni Rosenfeld. Proof of Activity: Extending Bitcoin's Proof of Work via Proof of Stake. *ACM SIGMETRICS Performance Evaluation Review*, 42(3):34–37, 2014.

[92] Karthikeyan Bhargavan, Antoine Delignat-Lavaud, Cédric Fournet, and et al. Formal Verification of Smart Contracts: Short Paper. In *Proceedings of the 2016 ACM Workshop on Programming Languages and Analysis for Security*, pages 91–96, 2016. ACM, Vienna Austria.

[93] Ronald L Rivest, Adi Shamir, and Leonard Adleman. A Method for Obtaining Digital Signatures and Public-Key Cryptosystems. *Communications of The ACM*, 21(2):120–126, 1978.

[94] Taher ElGamal. A public key cryptosystem and a signature scheme based on discrete logarithms. *IEEE Transactions on Information Theory*, 31(4):469–472, 1985.

[95] Victor S. Miller. The Weil Pairing, and Its Efficient Calculation. *Journal of Cryptology*, 17(4):235–261, 2004.

[96] M. Rabin. Digitalized Signatures and Public-Key Cryptosystems. *MIT/LCS/TR*, 212, 1979.

[97] Claus-Peter Schnorr. Efficient Identification and Signatures for Smart Cards. In *Conference on The Theory and Application of Cryptology*, pages 239–252. Springer, Houthalen, Belgium, 1989.

[98] Kaisa Nyberg and Rainer A Rueppel. A New Signature Scheme based on The DSA Giving Message Recovery. In *Proceedings of the 1st ACM conference on Computer and Communications Security*, pages 58–61, 1993. ACM, Fairfax Virginia USA.

[99] Kaisa Nyberg and Rainer A Rueppel. Message Recovery for Signature Schemes based on the Discrete Logarithm Problem. In *Workshop on the Theory and Application of of Cryptographic Techniques*, pages 182–193. Springer, Perugia, Italy, 1994.

[100] Ronald L Rivest, Martin E Hellman, John C Anderson, and John W Lyons. Responses to NIST's Proposal. *Communications of The ACM*, 35(7):41–54, 1992.

[101] François Morain. Building Cyclic Elliptic Curves Modulo Large Primes. In *Workshop on the Theory and Application of of Cryptographic Techniques*, pages 328–336. Springer, Brighton, UK, 1991.

[102] Shen Noether, Adam Mackenzie, et al. Ring Confidential Transactions. *Ledger*, 1:1–18, 2016.

[103] Don Johnson, Alfred Menezes, and Scott Vanstone. The Elliptic Curve Digital Signature Algorithm (ECDSA). *International Journal of Information Security*, 1(1):36–63, 2001.

[104] Benedikt Driessen, Axel Poschmann, and Christof Paar. Comparison of Innovative Signature Algorithms for WSNs. In *Proceedings of the first ACM conference on Wireless Network Security*, pages 30–35, 2008. ACM, Alexandria VA USA.

[105] Serge Vaudenay. The Security of DSA and ECDSA. In *International Workshop on Public Key Cryptography*, pages 309–323. Springer, Miami, FL, USA, 2003.

[106] Daniel RL Brown. The Exact Security of ECDSA. In *Advances in Elliptic Curve Cryptography*. Citeseer, London, UK, 2000.

[107] David Pointcheval and Jacques Stern. Security Proofs for Signature Schemes. In *International Conference on the Theory and Applications of Cryptographic Techniques*, pages 387–398. Springer, Saragossa, Spain, 1996.

[108] Ernest Brickell, David Pointcheval, Serge Vaudenay, and Moti Yung. Design Validations for Discrete Logarithm based Signature Schemes. In *International Workshop on Public Key Cryptography*, pages 276–292. Springer, Melbourne, VIC, Australia, 2000.

[109] Lidong Chen and Torben P Pedersen. New Group Signature Schemes. In *Workshop on the Theory and Application of of Cryptographic Techniques*, pages 171–181. Springer, Perugia, Italy, 1994.

[110] Sangjoon Park, Seungjoo Kim, and Dongho Won. ID-based Group Signature. *Electronics Letters*, 33(19):1616–1616, 1997.

[111] Wenbo Mao and Chae Hoon Lim. Cryptanalysis in Prime Order Subgroups of Z_n. In *International Conference on the Theory and Application of Cryptology and Information Security*, pages 214–226. Springer, Beijing, China, 1998.

[112] Yuh-Min Tseng and Jinn-Ke Jan. A Novel ID-based Group Signature. *Information sciences*, 120(1-4):131–141, 1999.

[113] Marc Joye. On the Difficulty Coalition-Resistance in Group Signature Schemes (II). *Technique Report*, 1999.

[114] Marc Joye, Seungjoo Kim, and Narn-Yih Lee. Cryptanalysis of Two Group Signature Schemes. In *International Workshop on Information Security*, pages 271–275. Springer, Kuala Lumpur, Malaysia, 1999.

[115] Jan Camenisch and Markus Stadler. Efficient group signature schemes for large groups. In *Annual International Cryptology Conference*, pages 410–424. Springer, 1997. California, USA.

[116] Jan Camenisch and Markus Michels. A Group Signature Scheme with Improved Efficiency. In *International Conference on the Theory and Application of Cryptology and Information Security*, pages 160–174. Springer, Beijing, China, 1998.

[117] Jan Camenisch and Markus Michels. A group Signature Scheme based on an RSA-Variant. *BRICS Report Series*, 5(27), 160–174. 1998.

[118] Seung Joo Kim, Sung Jun Park, and Dong Ho Won. Convertible Group Signatures. In *International Conference on the Theory and Application of Cryptology and Information Security*, pages 311–321. Springer, Kyongju, Korea, 1996.

[119] Claude Castelluccia. How to convert any id-based signature schemes into a group signature scheme. Doctoral dissertation, RR-4524, INRIA. 2002.

[120] Seungjoo Kim, Sangjoon Park, and Dongho Won. Group Signatures for Hierarchical Multigroups. In *International Workshop on Information Security*, pages 273–281. Springer, Tatsunokuchi, Ishikawa, Japan, 1997.

[121] Anna Lysyanskaya and Zulfikar Ramzan. Group Blind Digital Signatures: A Scalable Solution to Electronic Cash. In *International Conference on Financial Cryptography*, pages 184–197. Springer, Anguilla, 1998.

[122] Giuseppe Ateniese and Gene Tsudik. Some Open Issues and New Directions in Group Signatures. In *International Conference on Financial Cryptography*, pages 196–211. Springer, Anguilla, British West Indies, 1999.

[123] Xiaofeng Chen, Fangguo Zhang, and Kwangjo Kim. A New ID-based Group Signature Scheme from Bilinear Pairings. *IACR Cryptol. ePrint Arch.*, 2003:116, 2003.

[124] Victor K Wei, Tsz Hon Yuen, and Fangguo Zhang. Group Signature where Group Manager, Members and Open Authority are Identity-based. In *Australasian Conference on Information Security and Privacy*, pages 468–480. Springer, Melbourne, VIC, Australia, 2005.

[125] Toru Nakanishi, Hiroki Fujii, Yuta Hira, and Nobuo Funabiki. Revocable Group Signature Schemes with Constant Costs for Signing and Verifying. In *International Workshop on Public Key Cryptography*, pages 463–480. Springer, Orange County, Californi, 2009.

[126] Chun-I Fan, Ruei-Hau Hsu, and Mark Manulis. Group Signature with Constant Revocation Costs for Signers and Verifiers. In *International Conference on Cryptology and Network Security*, pages 214–233. Springer, Sanya, China, 2011.

[127] Benoît Libert, Thomas Peters, and Moti Yung. Scalable Group Signatures with Revocation. In *Annual International Conference on the Theory and Applications of Cryptographic Techniques*, pages 609–627. Springer, Cambridge, UK, 2012.

[128] Tsz Hon Yuen, Shi-feng Sun, Joseph K Liu, Man Ho Au, Muhammed F Esgin, Qingzhao Zhang, and Dawu Gu. RingCT 3.0 for Blockchain Confidential Transaction: Shorter Size and Stronger Security. In *International Conference on Financial Cryptography and Data Security*, pages 464–483. Springer, Kota Kinabalu, Malaysia, 2020.

[129] Mihir Bellare, Daniele Micciancio, and Bogdan Warinschi. Foundations of Group Signatures: Formal Definitions, Simplified Requirements, and a Construction based on General Assumptions. In *International Conference on the Theory and Applications of Cryptographic Techniques*, pages 614–629. Springer, Warsaw, Poland, 2003.

[130] Joseph K Liu, Victor K Wei, and Duncan S Wong. Linkable Spontaneous Anonymous Group Signature for Ad Hoc Groups. In *Australasian Conference on Information Security and Privacy*, pages 325–335. Springer, Sydney, Australia, 2004.

[131] Kazuo Ohta and Tatsuaki Okamoto. On Concrete Security Treatment of Signatures Derived from Identification. In *Annual International Cryptology Conference*, pages 354–369. Springer, Santa Barbara, California, USA, 1998.

[132] Ke Huang, Xiaosong Zhang, Yi Mu, Fatemeh Rezaeibagha, and Xiaojiang Du. Scalable and Redactable Blockchain with Update and Anonymity. *Information Sciences*, 546:25–41. 2020.

[133] Giuseppe Ateniese, Jan Camenisch, Marc Joye, and Gene Tsudik. A Practical and Provably Secure Coalition-Resistant Group Signature Scheme. In *Annual International Cryptology Conference*, pages 255–270. Springer, Santa Barbara, California, USA, 2000.

[134] Alexandra Boldyreva. Threshold Signatures, Multisignatures and Blind Signatures based on the Gap-Diffie-Hellman-Group Signature Scheme. In *International Workshop on Public Key Cryptography*, pages 31–46. Springer, Miami, FL, USA, 2003.

[135] Emmanuel Bresson, Jacques Stern, and Michael Szydlo. Threshold Ring Signatures and Applications to Ad-Hoc Groups. In *Annual International Cryptology Conference*, pages 465–480. Springer, Santa Barbara, CA, USA, 2002.

[136] Dennis YW Liu, Joseph K Liu, Yi Mu, Willy Susilo, and Duncan S Wong. Revocable Rng Signature. *Journal of Computer Science and Technology*, 22(6):785–794, 2007.

[137] Man Ho Au, Joseph K Liu, Willy Susilo, and Tsz Hon Yuen. Constant-size ID-based Linkable and Revocable-iff-Linked Ring Signature. In *International Conference on Cryptology in India*, pages 364–378. Springer, Kolkata, India, 2006.

[138] Patrick P Tsang, Victor K Wei, Tony K Chan, Man Ho Au, Joseph K Liu, and Duncan S Wong. Separable linkable threshold ring signatures. In *International Conference on Cryptology in India*, pages 384–398. Springer, Chennai, India, 2004.

[139] Joseph K Liu and Duncan S Wong. Linkable ring signatures: Security models and new schemes. In *International Conference on Computational Science and Its Applications*, pages 614–623. Springer, Singapore, 2005.

[140] Andrew S Tanenbaum and Maarten Van Steen. *Distributed Systems: Principles and Paradigms*. Prentice-Hall, 2007.

[141] Changsu Kim, Wang Tao, Namchul Shin, and Ki-Soo Kim. An Empirical Study of Customers Perceptions of Security and Trust in E-Payment Systems. *Electronic Commerce Research and Applications*, 9(1):84–95, 2010.

[142] Man Ho Au, Sherman SM Chow, Willy Susilo, and Patrick P Tsang. Short linkable ring signatures revisited. In *European Public Key Infrastructure Workshop*, pages 101–115. Springer, Turin, Italy, 2006.

[143] Joseph K Liu, Man Ho Au, Willy Susilo, and Jianying Zhou. Linkable Ring Signature with Unconditional Anonymity. *IEEE Transactions on Knowledge and Data Engineering*, 26(1):157–165, 2013.

[144] Tsz Hon Yuen, Joseph K Liu, Man Ho Au, Willy Susilo, and Jianying Zhou. Efficient Linkable and/or Threshold Ring Signature without Random Oracles. *The Computer Journal*, 56(4):407–421, 2013.

[145] Man Ho Au, Joseph K Liu, Willy Susilo, and Tsz Hon Yuen. Secure ID-based Linkable and Revocable-iff-Linked Ring Signature with Constant-Size Construction. *Theoretical Computer Science*, 469:1–14, 2013.

[146] Eiichiro Fujisaki. Sub-linear Size Traceable Ring Signatures without Random Oracles. In *Cryptographers Track at the RSA Conference*, pages 393–415. Springer, San Francisco, CA, USA, 2011.

[147] Eiichiro Fujisaki and Koutarou Suzuki. Traceable Ring Signature. In *International Workshop on Public Key Cryptography*, pages 181–200. Springer, Beijing, P.R. China IACR, 2007.

[148] Xiaofeng Chen, Fangguo Zhang, Haibo Tian, Baodian Wei, and Kwangjo Kim. Key-Exposure Free Chameleon Hashing and Signatures Based on Discrete Logarithm Systems. *IACR Cryptology ePrint Archive*, 2009:35, 2009.

[149] Xiaofeng Chen, Fangguo Zhang, and Kwangjo Kim. Chameleon Hashing without Key Exposure. In *International Conference on Information Security*, pages 87–98. Springer, Palo Alto, CA, USA, 2004.

[150] Ke Huang, Xiaosong Zhang, Yi Mu, Xiaofen Wang, Guomin Yang, Xiaojiang Du, Qi Xia, Fatemeh Rezaeibagha, and Mohsen Guizani. Building Redactable Consortium Blockchain for Industrial Internet-of-Things. In *IEEE Transactions on Industrial Informatics*, 2019.

[151] Shafi Goldwasser and Silvio Micali. Probabilistic Encryption &: How to Play Mental Poker Keeping Secret All Partial Information. In *Proceedings of the Fourteenth Annual ACM Symposium on Theory of Computing*, ACM, New York, USA, pages 365–377, 1982.

[152] Shafi Goldwasser and Silvio Micali. Probabilistic Encryption. *Journal of Computer and System Sciences*, 28(2):270–299, 1984.

[153] Moni Naor and Moti Yung. Public-Key Cryptosystems Provably Secure against Chosen Ciphertext Attacks. In *Proceedings of The Twenty-Second Annual ACM Symposium on Theory of Computing*, ACM, Baltimore Maryland USA, pages 427–437, 1990.

[154] Darrel Hankerson, Alfred J Menezes, and Scott Vanstone. *Guide to Elliptic Curve Cryptography*. Springer Science & Business Media, 2006.

[155] Torben Pryds Pedersen. Non-interactive and information-theoretic secure verifiable secret sharing. In *Annual International Cryptology Conference*, pages 129–140. Springer, Barbara, California, USA, 1991.

[156] Dan Boneh and Matt Franklin. Identity-based Encryption from the Weil Pairing. In *Annual International Cryptology Conference*, pages 213–229. Springer, Santa Barbara, CA, USA, 2001.

[157] Adi Shamir. Identity-based Cryptosystems and Signature Schemes. In *Workshop on The Theory and Application of Cryptographic Techniques*, pages 47–53. Springer, Paris, France, 1984.

[158] James Backhouse, Carol Hsu, and Aidan McDonnell. Toward Public-key Infrastructure Interoperability. *Communications of The ACM*, 46(6):98–100, 2003.

[159] Dan Boneh and Xavier Boyen. Secure Identity based Encryption without Random Oracles. In *Annual International Cryptology Conference*, pages 443–459. Springer, Santa Barbara, CA, 2004.

[160] R Canetti. Efficient Selectiveid Identity based Encryption without Random Oracles. In *Advances in Cryptology-EUROCRYPT'04*, pages 207–222. Springer-Verlag, Interlaken, Switzerland, 2004.

[161] Brent Waters. Efficient Identity-based Encryption without Random Oracles. In *Annual International Conference on the Theory and Applications of Cryptographic Techniques*, pages 114–127. Springer, Aarhus, Denmark, 2005.

[162] Craig Gentry. Practical Identity-based Encryption without Random Oracles. In *Annual International Conference on the Theory and Applications of Cryptographic Techniques*, pages 445–464. Springer, St. Petersburg, Russia, 2006.

[163] Shushan Zhao, Akshai Aggarwal, Richard Frost, and Xiaole Bai. A Survey of Applications of Identity-based Cryptography in Mobile Ad-Hoc Networks. *IEEE Communications Surveys & Tutorials*, 14(2):380–400, 2011.

[164] Beini Zhou, Hui Li, and Li Xu. An Authentication Scheme using Identity-based Encryption & Blockchain. In *2018 IEEE Symposium on Computers and Communications (ISCC)*, pages 00556–00561. IEEE, Natal, Brazil, 2018.

[165] Jingting Xue, Chunxiang Xu, Jining Zhao, and Jianfeng Ma. Identity-based Public Auditing for Cloud Storage Systems against Malicious Auditors via Blockchain. *Science China Information Sciences*, 62(3):32104, 2019.

[166] Shu Yun Lim, Pascal Tankam Fotsing, Abdullah Almasri, Omar Musa, Miss Laiha Mat Kiah, Tan Fong Ang, and Reza Ismail. Blockchain Technology The Identity Management and Authentication Service Disruptor: A Survey. *International Journal on Advanced Science, Engineering and Information Technology*, 8(4-2):1735–1745, 2018.

[167] Peter Gutmann. PKI: It's Not Dead, Just Resting. *Computer*, 35(8):41–49, 2002.

[168] Eiichiro Fujisaki and Tatsuaki Okamoto. Secure Integration of Asymmetric and Symmetric Encryption Schemes. In *Annual International Cryptology Conference*, pages 537–554. Springer, Santa Barbara, CA, 1999.

[169] Ronald Cramer and Victor Shoup. A Practical Public Key Cryptosystem Provably Secure Against Ddaptive Chosen Ciphertext Attack. In *Annual International Cryptology Conference*, pages 13–25. Springer, Santa Barbara, CA, USA, 1998.

[170] Yuliang Zheng. Digital Signcryption or How To Achieve Cost (Signature & Encryption)? Cost (Signature)+ Cost (Encryption). In *Annual International Cryptology Conference*, pages 165–179. Springer, Santa Barbara, California, 1997.

[171] Joonsang Baek, Ron Steinfeld, and Yuliang Zheng. Formal Proofs for The Security of Signcryption. In *International Workshop on Practice Theory in Public Key Cryptosystems: Public Key Cryptography*. Springer, Paris, France, 2002.

[172] Feng Bao and Robert H. Deng. A Signcryption Scheme with Signature Directly Verifiable by Public Key. *Proc Pck98 Feb*, 1431:55–59, 1998.

[173] Chandana Gamage, Jussipekka Leiwo, and Yuliang Zheng. Encrypted Message Authentication by Firewalls. In *International Workshop on Public Key Cryptography*. Springer, Kamakura, Japan, 1999.

[174] John Malonelee and Wenbo Mao. Two Birds One Stone: Signcryption Using RSA. In *RSA Conference on the Cryptographers Track*. Springer, San Francisco, CA, USA, 2003.

[175] Liqun Chen and John Malone-Lee. Improved Identity-Based Signcryption. *Lecture Notes in Computer Science*, 2004(1):362–379, 2005.

[176] Xavier Boyen. Multipurpose Identity-based Signcryption. In *Annual International Cryptology Conference*, pages 383–399. Springer, Santa Barbara, CA, USA, 2003.

[177] Sherman SM Chow, Siu-Ming Yiu, Lucas CK Hui, and KP Chow. Efficient Forward and Provably Secure ID-based Signcryption Scheme with

Public Verifiability and Public Ciphertext Authenticity. In *International Conference on Information Security and Cryptology*, pages 352–369. Springer, Seoul, Korea, 2003.

[178] B Libert and J. J. Quisquater. A new identity-based signcryption scheme from pairings. In *IEEE Information Theory Workshop*, IEEE, Paris, France, pages 155–158, 2003.

[179] Tsz Hon Yuen and Victor K. Wei. Fast and Proven Secure Blind Identity-Based Signcryption from Pairings. In *International Conference on Topics in Cryptology*. ACM, San Francisco, CA, USA, 2005.

[180] Paulo S. L. M. Barreto, Benoit Libert, Noel Mccullagh, and Jean Jacques Quisquater. Efficient and Provably-Secure Identity-Based Signatures and Signcryption from Bilinear Maps. In *International Conference on Theory Application of Cryptology Information Security*. Springer, San Francisco, CA, USA, 2005.

[181] Fagen Li, Muhammad Khurram Khan, Khaled Alghathbar, and Tsuyoshi Takagi. Identity-based Online/Offline Signcryption for Low Power Devices. *Journal of Network Computer Applications*, 35(1):340–347, 2012.

[182] Joseph K Liu, Joonsang Baek, and Jianying Zhou. Online/offline Identity-based Signcryption Revisited. In *International Conference on Information Security and Cryptology*, pages 36–51. Springer, Shanghai, China, 2010.

[183] S. Sharmila Deva Selvi, S. Sree Vivek, and C. Pandu Rangan. Identity Based Online/Offline Encryption and Signcryption Schemes Revisited. In *International Conference on Security Aspects in Information Technology*. Springer, Haldia, India, 2011.

[184] Jianchang Lai, Yi Mu, and Fuchun Guo. Efficient Identity-based Online/Offline Encryption and Signcryption with Short Ciphertext. *International Journal of Information Security*, 16(3):299–311, 2017.

[185] Zhengping Jin, Qiaoyan Wen, and Hongzhen Du. An Improved Semantically-Secure Identity-based Signcryption Scheme in The Standard Model. *Computers Electrical Engineering*, 36(3):545–552, 2010.

[186] Jin Li, Xiaofeng Chen, Mingqiang Li, Jingwei Li, Patrick PC Lee, and Wenjing Lou. Secure Deduplication with Efficient and Reliable Convergent Key Management. *IEEE Transactions on Parallel and Distributed Systems*, 25(6):1615–1625, 2013.

[187] Xiangxue Li, Haifeng Qian, Weng Jian, and Yu Yu. Fully Secure Identity-based Signcryption Scheme with Shorter Signcryptext in The Standard Model. *Mathematical Computer Modelling*, 57(3-4):503–511, 2013.

[188] S Sharmila Deva Selvi, S Sree Vivek, Dhinakaran Vinayagamurthy, and C Pandu Rangan. ID based Signcryption Scheme in Standard Model. In *International Conference on Provable Security*, pages 35–52. Springer, Chengdu, China, 2012.

[189] Arijit Karati, Sk Hafizul Islam, G. P. Biswas, and et al. Provably Secure Identity-based Signcryption Scheme for Crowdsourced Industrial Internet of Things Environments. *IEEE Internet of Things Journal*, PP(99):1–1, 2017.

[190] Ke Huang, Xiaosong Zhang, Xiaofen Wang, and et al. HUCDO: A Hybrid User-centric Data Outsourcing Scheme. *ACM Transactions on Cyber-Physical Systems*, 4(3):1–23, 2020.

[191] Fatemeh Rezaeibagha, Mu Yi, Shiwei Zhang, and Xiaofen Wang. Provably secure homomorphic signcryption. In *International Conference on Provable Security*, Springer, Xi'an, China, 2017.

[192] Michel Abdalla, Mihir Bellare, and Phillip Rogaway. DHAES: An Encryption Scheme Based on the Diffie-Hellman Problem. *IACR Cryptol. ePrint Arch.*, 1999:7, 1999.

[193] Michael O Rabin. Digital Signatures and Public-Key Functions as Intractable as Factorization. *MIT Laboratory for Computer Science Technical Report*, 1979.

[194] Manuel Blum and Shafi Goldwasser. An Efficient Probabilistic Public-Key Encryption Scheme Which Hides All Partial Information. In *Workshop on The Theory and Application of Cryptographic Techniques*, pages 289–299. Springer, Paris, France, 1984.

[195] Yuliang Zheng and Jennifer Seberry. Practical Approaches to Attaining Security Against Adaptively Chosen Ciphertext Attacks. In *Annual International Cryptology Conference*, pages 292–304. Springer, Santa Barbara, California, USA, 1992.

[196] Ronald Cramer and Victor Shoup. Universal Hash Proofs and A Paradigm for Adaptive Chosen Ciphertext Secure Public-Key Encryption. In *International Conference on the Theory and Applications of Cryptographic Techniques*, pages 45–64. Springer, Amsterdam, 2002.

[197] Ran Canetti, Shai Halevi, and Jonathan Katz. Chosen-ciphertext security from identity-based encryption. In *International Conference on the Theory and Applications of Cryptographic Techniques*, pages 207–222. Springer, 2004.

[198] Uriel Feige, Amos Fiat, and Adi Shamir. Zero-Knowledge Proofs of Identity. *Journal of Cryptology*, 1(2):77–94, 1988.

[199] Zvi Galil, Stuart Haber, and Moti Yung. Symmetric Public-Key Encryption. In *Conference on the Theory and Application of Cryptographic Techniques*, pages 128–137. Springer, Linz, Austria, 1985.

[200] Manuel Blum, Paul Feldman, and Silvio Micali. Non-Interactive Zero-Knowledge and Its Applications. In *Providing Sound Foundations for Cryptography: On the Work of Shafi Goldwasser and Silvio Micali*, ACM, pages 329–349. 2019.

[201] Oded Goldreich, Shafi Goldwasser, and Silvio Micali. On The Cryptographic Applications of Random Functions. In *Workshop on the Theory and Application of Cryptographic Techniques*, pages 276–288. Springer, Paris, France, 1984.

[202] Oded Goldreich, Shafi Goldwasser, and Silvio Micali. How to Construct Random Functions. In *Providing Sound Foundations for Cryptography: On the Work of Shafi Goldwasser and Silvio Micali*, ACM, pages 241–264. 2019.

[203] Victor Shoup. OAEP Reconsidered. In *Annual International Cryptology Conference*, pages 239–259. Springer, Santa Barbara, California, USA, 2001.

[204] Eiichiro Fujisaki, Tatsuaki Okamoto, David Pointcheval, and Jacques Stern. RSA-OAEP is Secure under the RSA Assumption. In *Annual International Cryptology Conference*, pages 260–274. Springer, Santa Barbara, CA, USA, 2001.

[205] Eiichiro Fujisaki and Tatsuaki Okamoto. How to Enhance the Security of Public-key Encryption at Minimum Cost. In *International Workshop on Public Key Cryptography*, pages 53–68. Springer, 1999.

[206] David Pointcheval. Chosen-Ciphertext Security for any One-Way Cryptosystem. In *International Workshop on Public Key Cryptography*, pages 129–146. Springer, Pacifico Yokohama, Japan, 2000.

[207] Tatsuaki Okamoto and David Pointcheval. REACT: Rapid Enhanced-Security Asymmetric Cryptosystem Transform. In *Cryptographers Track at the RSA Conference*, pages 159–174. Springer, San Francisco, CA, USA, 2001.

[208] Ronald Cramer and Victor Shoup. Design and Analysis of Practical Public-key Encryption Schemes Secure against Adaptive Chosen Ciphertext Attack. *SIAM Journal on Computing*, 33(1):167–226, 2003.

[209] Moni Naor and Moti Yung. Universal One-Way Hash Functions and Their Cryptographic Applications. In *Proceedings of The Twenty-First Annual ACM Symposium on Theory of Computing*, pages 33–43, 1989.

[210] Pascal Paillier. Public-Key Cryptosystems based on Composite Degree Residuosity Classes. In *International Conference on The Theory and Applications of Cryptographic Techniques*, pages 223–238. Springer, Prague, Czech Republic, 1999.

[211] Kaoru Kurosawa and Yvo Desmedt. A New Paradigm of Hybrid Encryption Scheme. In *Annual International Cryptology Conference*, pages 426–442. Springer, Santa Barbara, CA, USA, 2004.

[212] Eike Kiltz. Chosen-Ciphertext Secure Key-Encapsulation based on Gap Hashed Diffie-Hellman. In *International Workshop on Public Key Cryptography*, pages 282–297. Springer, Beijing, China, 2007.

[213] Dan Boneh, Antoine Joux, and Phong Q Nguyen. Why Textbook ElGamal and RSA Encryption Are Insecure. In *International Conference on the Theory and Application of Cryptology and Information Security*, pages 30–43. Springer, Kyoto, Japan, 2000.

[214] Marc Fischlin. *Trapdoor Commitment Schemes and Their Applications.* PhD thesis, Citeseer, 2001.

[215] Giuseppe Ateniese, Bernardo Magri, Daniele Venturi, and Ewerton Andrade. Redactable Blockchain–or–Rewriting History in Bitcoin and Friends. In *2017 IEEE European Symposium on Security and Privacy (EuroS&P)*, pages 111–126. IEEE, Paris, France, 2017.

[216] Dominic Deuber, Bernardo Magri, and Sri Aravinda Krishnan Thyagarajan. Redactable Blockchain in The Permissionless Setting. *arXiv preprint arXiv:1901.03206*, 2019.

[217] Richard Lumb, D Treat, and O Jelf. Editing the Uneditable Blockchain-Why Distributed Ledger Technology must Adapt to an Imperfect World. *https://newsroom. accenture. com/content/1101/files/Cross-FSBC.pdf [accesssed October 2017]*, 2016.

[218] Hugo Mario Krawczyk and Tal D Rabin. Chameleon Hashing and Signatures, August 22, 2000. US Patent 6,108,783.

[219] Giuseppe Ateniese and Breno de Medeiros. Identity-based chameleon hash and applications. In *International Conference on Financial Cryptography*, pages 164–180. Springer, 2004.

[220] Stephan Krenn, Henrich C Pöhls, Kai Samelin, and Daniel Slamanig. Chameleon-hashes with dual long-term trapdoors and their applications. In *International Conference on Cryptology in Africa*, pages 11–32. Springer, 2018.

[221] Junzuo Lai, Xuhua Ding, and Yongdong Wu. Accountable Trapdoor Sanitizable Signatures. In *International Conference on Information*

Security Practice and Experience, pages 117–131. Springer, Lanzhou, China, 2013.

[222] Xiaofeng Chen, Haibo Tian, Fangguo Zhang, and Yong Ding. Comments and Improvements on Key-Exposure Free Chameleon Hashing based on Factoring. In *International Conference on Information Security and Cryptology*, pages 415–426. Springer, Seoul, Korea, 2010.

[223] Mihir Bellare and Todor Ristov. A Characterization of Chameleon Hash Functions and New, Efficient Designs. *Journal of Cryptology*, 27(4):799–823, 2014.

[224] Ke Huang, Xiaosong Zhang, Yi Mu, Fatemeh Rezaeibagha, Xiaojiang Du, and Nadra Guizani. Achieving intelligent trust-layer for iot via self-redactable blockchain. *IEEE Transactions on Industrial Informatics*, 16 (4):2677–2686, 2019.

[225] Jens Groth and Amit Sahai. Efficient Non-Interactive Proof Systems for Bilinear Groups. In *Annual International Conference on the Theory and Applications of Cryptographic Techniques*, pages 415–432. Springer, Istanbul, Turkey, 2008.

[226] Jens Groth. On The Size of Pairing-based Non-Interactive Arguments. In *Annual International Conference on The Theory and Applications of Cryptographic Techniques*, pages 305–326. Springer, Vienna, Austria, 2016.

[227] Matteo Campanelli, Rosario Gennaro, Steven Goldfeder, and Luca Nizzardo. Zero-knowledge Contingent Payments Revisited: Attacks and Payments for Services. In *Proceedings of the 2017 ACM SIGSAC Conference on Computer and Communications Security*, ACM, Dallas Texas USA, pages 229–243, 2017.

[228] Bryan Parno, Jon Howell, Craig Gentry, and Mariana Raykova. Pinocchio: Nearly Practical Verifiable Computation. In *2013 IEEE Symposium on Security and Privacy*, pages 238–252. IEEE, San Francisco, California, 2013.

[229] Amos Fiat and Adi Shamir. How to Prove Yourself: Practical Solutions to Identification and Signature Problems. In *Conference on The Theory and Application of Cryptographic Techniques*, pages 186–194. Springer, 1986.

[230] Josh Benaloh and Michael De Mare. One-Way Accumulators: A Decentralized Alternative to Digital Signatures. In *Workshop on the Theory and Application of of Cryptographic Techniques*, pages 274–285. Springer, Lofthus, Norway, 1993.

[231] Christina Garman, Matthew Green, Ian Miers, and Aviel D Rubin. Rational Zero: Economic Security for Zerocoin with Everlasting Anonymity. In *International Conference on Financial Cryptography and Data Security*, pages 140–155. Springer, Barbados, 2014.

[232] George Danezis, Cedric Fournet, Markulf Kohlweiss, and Bryan Parno. Pinocchio Coin: Building Zerocoin from a Succinct Pairing-based proof System. In *Proceedings of the First ACM Workshop on Language Support for Privacy-enhancing Technologies*, pages 27–30, 2013, Berlin Germany.

[233] Benedikt Bünz, Shashank Agrawal, Mahdi Zamani, and Dan Boneh. Zether: Towards Privacy in a Smart Contract World. *IACR Cryptol. ePrint Arch.*, 2019:191, 2019.

[234] Gregory Maxwell. CoinJoin: Bitcoin Privacy for the Real World. In *Post on Bitcoin Forum*, 2013. https://bitcointalk.org/index.php?topic=279249.0.

[235] Eli Ben-Sasson, Alessandro Chiesa, Daniel Genkin, Eran Tromer, and Madars Virza. SNARKs for C: Verifying Program Executions Succinctly and in Zero Knowledge. In *Annual Cryptology Conference*, pages 90–108. Springer, Santa Barbara, CA, USA, 2013.

[236] Eli Ben-Sasson, Alessandro Chiesa, Eran Tromer, and Madars Virza. Succinct Non-Interactive Zero Knowledge for a Von Neumann Architecture. In *23rd {USENIX} Security Symposium ({USENIX} Security 14)*, USENIX, San, Diego, CA, pages 781–796, 2014.

[237] Tiacheng Xie, Jiaheng Zhang, Yupeng Zhang, Charalampos Papamanthou, and Dawn Song. Libra: Succinct Zero-Knowledge Proofs with Optimal Prover Computation. In *Annual International Cryptology Conference*, pages 733–764. Springer, Santa Barbara, CA, USA, 2019.

[238] Scott Ames, Carmit Hazay, Yuval Ishai, and Muthuramakrishnan Venkitasubramaniam. Ligero: Lightweight Sublinear Arguments without a Trusted Setup. In *Proceedings of the 2017 ACM SIGSAC Conference on Computer and Communications Security*, ACM, Dallas Texas USA, pages 2087–2104, 2017.

[239] Benedikt Bünz, Jonathan Bootle, Dan Boneh, Andrew Poelstra, Pieter Wuille, and Greg Maxwell. Bulletproofs: Short Proofs for Confidential Transactions and More. In *2018 IEEE Symposium on Security and Privacy (SP)*, pages 315–334. IEEE, San Francisco, CA, 2018.

[240] Riad S Wahby, Ioanna Tzialla, Abhi Shelat, Justin Thaler, and Michael Walfish. Doubly-Efficient zkSNARKs without Trusted Setup. In *2018 IEEE Symposium on Security and Privacy (SP)*, pages 926–943. IEEE, San Francisco, CA, USA, 2018.

[241] Eli Ben-Sasson, Iddo Bentov, Yinon Horesh, and Michael Riabzev. Scalable, Transparent, and Post-Quantum Secure Computational Integrity. *IACR Cryptol. ePrint Arch.*, 2018:46, 2018.

[242] Eli Ben-Sasson, Alessandro Chiesa, Michael Riabzev, Nicholas Spooner, Madars Virza, and Nicholas P Ward. Aurora: Transparent Succinct Arguments for R1CS. In *Annual International Conference on The Theory and Applications of Cryptographic Techniques*, pages 103–128. Springer, Darmstadt, Germany, 2019.

[243] Benjamin Braun, Ariel J Feldman, Zuocheng Ren, Srinath Setty, Andrew J Blumberg, and Michael Walfish. Verifying Computations with State. In *Proceedings of the Twenty-Fourth ACM Symposium on Operating Systems Principles*, Farminton Pennsylvania, pages 341–357, 2013.

[244] Ahmed E Kosba, Dimitrios Papadopoulos, Charalampos Papamanthou, Mahmoud F Sayed, Elaine Shi, and Nikos Triandopoulos. {TRUESET}: Faster Verifiable Set Computations. In *23rd {USENIX} Security Symposium ({USENIX} Security 14)*, USENIX, San, Diego, CA, pages 765–780, 2014.

[245] Michael Backes, Manuel Barbosa, Dario Fiore, and Raphael M Reischuk. ADSNARK: Nearly Practical and Privacy-Preserving Proofs on Authenticated Data. In *2015 IEEE Symposium on Security and Privacy*, pages 271–286. IEEE, San Jose, CA, USA, 2015.

[246] Alessandro Chiesa, Eran Tromer, and Madars Virza. Cluster Computing in Zero Knowledge. In *Annual International Conference on the Theory and Applications of Cryptographic Techniques*, pages 371–403. Springer, Sofia, Bulgaria, 2015.

[247] Rosario Gennaro, Craig Gentry, Bryan Parno, and Mariana Raykova. Quadratic Span Programs and Succinct NIZKs without PCPs. In *Annual International Conference on the Theory and Applications of Cryptographic Techniques*, pages 626–645. Springer, Athens, Greece, 2013.

[248] Jens Groth. Simulation-sound NIZK Proofs for A Practical Language and Constant Size Group Signatures. In *International Conference on the Theory and Application of Cryptology and Information Security*, pages 444–459. Springer, Shanghai, China, 2006.

[249] Nir Bitansky, Alessandro Chiesa, Yuval Ishai, Omer Paneth, and Rafail Ostrovsky. Succinct non-interactive arguments via linear interactive proofs. In *Theory of Cryptography Conference*, pages 315–333. Springer, 2013.

[250] Jens Groth. Short Pairing-based Non-Interactive Zero-Knowledge Arguments. In *International Conference on the Theory and Application of Cryptology and Information Security*, pages 321–340. Springer, Seoul, Korea, 2010.

[251] Helger Lipmaa. Progression-Free Sets and Sublinear Pairing-based Non-interactive Zero-Knowledge Arguments. In *Theory of Cryptography Conference*, pages 169–189. Springer, Taormina, Sicily, Italy, 2012.

[252] Helger Lipmaa. Succinct Non-Interactive Zero Knowledge Arguments from Span Programs and Linear Error-Correcting Codes. In *International Conference on the Theory and Application of Cryptology and Information Security*, pages 41–60. Springer, Bengaluru, India, 2013.

[253] Rosario Gennaro. Multi-Trapdoor Commitments and Their Applications to Proofs of Knowledge Secure under Concurrent Man-in-The-Middle Attacks. In *Annual International Cryptology Conference*, pages 220–236. Springer, Santa Barbara, CA, USA, 2004.

[254] Vitalik Buterin. Quadratic Arithmetic Programs: from Zero to Hero. *https://medium.com/@VitalikButerin/quadratic-arithmetic-programs-from-zero-to-hero-f6d558cea649 (accessed Jan 14, 2021)*.

[255] Aritra Banerjee, Michael Clear, and Hitesh Tewari. Demystifying the Role of zk-SNARKs in Zcash. *arXiv preprint arXiv:2008.00881*, 2020.

[256] Jean-Paul Berrut and Lloyd N Trefethen. Barycentric Lagrange Interpolation. *SIAM Review*, 46(3):501–517, 2004.

[257] Dana Moshkovitz. An Alternative Proof of The Schwartz-Zippel Lemma. In *Electronic Colloquium on Computational Complexity (ECCC)*, volume 17, page 34, 2010. https://eccc.weizmann.ac.il/report/2010/096/revision/1/download/

[258] S. Bowe. Pay-to-sudokus. 2017. https://electriccoin.co/blog/pay-to-sudoku-revisited/

[259] Nir Kshetri. Can Blockchain Strengthen the Internet of Things? *IT Professional*, 19(4):68–72, 2017.

[260] Guy Zyskind, Oz Nathan, et al. Decentralizing Privacy: Using bBlockchain to Protect Personal Data. In *2015 IEEE Security and Privacy Workshops*, pages 180–184. IEEE, San Jose, California, 2015.

[261] Tomaso Aste, Paolo Tasca, and Tiziana Di Matteo. Blockchain Technologies: The Foreseeable Impact on Society and Industry. *Computer*, 50(9):18–28, 2017.

[262] Man Ho Allen Au. Contribution to Privacy-Preserving Cryptographic Techniques. *https://ro.uow.edu.au/cgi/viewcontent.cgi?article=1826&context=theses (accessed Jan 14, 2021)*.

[263] Gilles Brassard, David Chaum, and Claude Crépeau. Minimum Disclosure Proofs of Knowledge. *Journal of Computer and System Cciences*, 37(2):156–189, 1988.

[264] Uriel Feige and Adi Shamir. Zero Knowledge Proofs of Knowledge in Two Rounds. In *Conference on the Theory and Application of Cryptology*, pages 526–544. Springer, Santa Barbara, 1989.

[265] Gilles Brassard, Claude Crépeau, and Moti Yung. Constant-Round Perfect Zero-Knowledge Computationally Convincing Protocols. *Theoretical Computer Science*, 84(1):23–52, 1991.

[266] Mihir Bellare, Silvio Micali, and Rafail Ostrovsky. Perfect Zero-Knowledge in Constant Rounds. In *Proceedings of the Twenty-Second Annual ACM Symposium on Theory of Computing*, ACM, Baltimore Maryland USA, pages 482–493, 1990.

[267] Giovanni Di Crescenzo and Rafail Ostrovsky. On Concurrent Zero-Knowledge with Pre-Processing. In *Annual International Cryptology Conference*, pages 485–502. Springer, Santa Barbara, California, 1999.

[268] Ivan Damgård. Efficient Concurrent Zero-Knowledge in The Auxiliary String Model. In *International Conference on the Theory and Applications of Cryptographic Techniques*, pages 418–430. Springer, Bruges, Belgium, 2000.

[269] Ran Canetti, Oded Goldreich, Shafi Goldwasser, and Silvio Micali. Resettable Zero-Knowledge. In *Proceedings of the Thirty-Second Annual ACM Symposium on Theory of Computing*, pages 235–244, 2000.

[270] Rosario Gennaro, Shai Halevi, and Tal Rabin. Secure Hash-and-Sign Signatures without The Random Oracle. In *International Conference on the Theory and Applications of Cryptographic Techniques*, pages 123–139. Springer, Prague, Czech Republic, 1999.

[271] Ronald Cramer and Victor Shoup. Signature sSchemes based on the Strong RSA Assumption. *ACM Transactions on Information and System Security (TISSEC)*, 3(3):161–185, 2000.

[272] Kondapally Ashritha, M Sindhu, and KV Lakshmy. Redactable Blockchain using Enhanced Chameleon Hash Function. In *2019 5th International Conference on Advanced Computing & Communication Systems (ICACCS)*, pages 323–328. IEEE, 2019. Sri Eshwar College of Engineering, India.

[273] David Derler, Kai Samelin, Daniel Slamanig, and Christoph Striecks. Fine-Grained and Controlled Rewriting in Blockchains: Chameleon-Hashing Gone Attribute-Based. *IACR Cryptol. ePrint Arch.*, 2019:406, 2019.

[274] Melissa Chase, Markulf Kohlweiss, Anna Lysyanskaya, and Sarah Meiklejohn. Malleable Signatures: New Definitions and Delegatable Anonymous Credentials. In *2014 IEEE 27th Computer Security Foundations Symposium*, pages 199–213. IEEE, 2014.

[275] Henrich C Pöhls, Stefan Peters, Kai Samelin, Joachim Posegga, and Hermann de Meer. Malleable Signatures for Resource Constrained Platforms. In *IFIP International Workshop on Information Security Theory and Practices*, pages 18–33. Springer, 2013.

[276] David Chaum and Hans Van Antwerpen. Undeniable Signatures. In *Conference on the Theory and Application of Cryptology*, pages 212–216. Springer, 1989.

[277] Giuseppe Ateniese, Daniel H Chou, Breno De Medeiros, and Gene Tsudik. Sanitizable Signatures. In *European Symposium on Research in Computer Security*, pages 159–177. Springer, 2005.

[278] Christina Brzuska, Marc Fischlin, Tobias Freudenreich, Anja Lehmann, Marcus Page, Jakob Schelbert, Dominique Schröder, and Florian Volk. Security of Sanitizable Signatures Revisited. In *International Workshop on Public Key Cryptography*, pages 317–336. Springer, 2009.

[279] Christina Brzuska, Marc Fischlin, Anja Lehmann, and Dominique Schröder. Unlinkability of Sanitizable Signatures. In *International Workshop on Public Key Cryptography*, pages 444–461. Springer, 2010.

[280] Sébastien Canard, Amandine Jambert, and Roch Lescuyer. Sanitizable Signatures with Several Signers and Sanitizers. In *International Conference on Cryptology in Africa*, pages 35–52. Springer, 2012.

[281] Ee-Chien Chang, Chee Liang Lim, and Jia Xu. Short Redactable Signatures using Random Trees. In *Cryptographers Track at the RSA Conference*, pages 133–147. Springer, 2009.

[282] Henrich C Pöhls and Kai Samelin. Accountable Redactable Signatures. In *2015 10th International Conference on Availability, Reliability and Security*, pages 60–69. IEEE, 2015.

[283] Roman Matzutt, Martin Henze, Jan Henrik Ziegeldorf, Jens Hiller, and Klaus Wehrle. Thwarting Unwanted Blockchain Content Insertion. In *2018 IEEE International Conference on Cloud Engineering (IC2E)*, pages 364–370. IEEE, 2018.

[284] Ivan Puddu, Alexandra Dmitrienko, and Srdjan Capkun. μchain: How to Forget without Hard Forks. *IACR Cryptology ePrint Archive*, 2017:106, 2017.

[285] Yevgeniy Dodis, Aggelos Kiayias, Antonio Nicolosi, and Victor Shoup. Anonymous Identification in Ad Hoc Groups. In *International Conference on the Theory and Applications of Cryptographic Techniques*, pages 609–626. Springer, 2004.

[286] Jan Camenisch and Anna Lysyanskaya. Dynamic Accumulators and Application to Efficient Revocation of Aonymous Credentials. In *Annual International Cryptology Conference*, pages 61–76. Springer, 2002.

[287] Niko Barić and Birgit Pfitzmann. Collision-Free Accumulators and Fail-Stop Signature Schemes without Trees. In *International Conference on the Theory and Applications of Cryptographic Techniques*, pages 480–494. Springer, 1997.

[288] Burton H Bloom. Space/Time Trade-Offs in Hash Coding with Allowable Errors. *Communications of the ACM*, 13(7):422–426, 1970.

[289] James K. Mullin. Optimal Semijoins for Distributed Database Systems. *IEEE Transactions on Software Engineering*, 16(5):558–560, 1990.

[290] Zhe Li and Kenneth A Ross. Perf Join: An Alternative to Two-Way Semijoin and Bloomjoin. In *Proceedings of the Fourth International Conference on Information and Knowledge Management*, pages 137–144, 1995.

[291] Moni Naor and Eylon Yogev. Bloom Filters in Adversarial Environments. In *Annual Cryptology Conference*, pages 565–584. Springer, 2015.

[292] Rafael Pass and Abhi Shelat. Micropayments for Decentralized Currencies. In *Proceedings of the 22nd ACM SIGSAC Conference on Computer and Communications Security*, pages 207–218, 2015. ACM, Denver Colorado USA.

[293] Bank St Louis, Joseph R Coyne, Chairman Stephen, H Axilrod, John M Denkler, and Michael J Prell. Board of Governors of the Federal Reserve System. 1979. http://citeseerx.ist.psu.edu/viewdoc/summary?doi=10.1.1.366.1255.

[294] Silvio Micali and Ronald L Rivest. Micropayments Revisited. In *Cryptographers Track at the RSA Conference*, pages 149–163. Springer, 2002.

[295] Alessandro Chiesa, Matthew Green, Jingcheng Liu, and et al. Decentralized Anonymous Micropayments. In *Annual International Conference on the Theory and Applications of Cryptographic Techniques*, pages 609–642. Springer, 2017.

[296] Ghada Almashaqbeh, Allison Bishop, and Justin Cappos. Microcash: Practical Concurrent Processing of Micropayments. In *International Conference on Financial Cryptography and Data Security*, pages 227–244. Springer, 2020.

[297] Manny Trillo. Stress Test Prepares Visanet for The Most Wonderful Time of The Year. *https://www.visa.com/blogarchives/us/category/visanet-2/index.html*, 2013.

[298] Ferenc Béres, Istvan Andras Seres, and András A Benczúr. A Cryptoeconomic Traffic Analysis of Bitcoins Lightning Network. *arXiv preprint arXiv:1911.09432*, 2019.

[299] Andrew Miller, Iddo Bentov, Surya Bakshi, Ranjit Kumaresan, and Patrick McCorry. Sprites and State Channels: Payment Networks That Go Faster Than Lightning. In *International Conference on Financial Cryptography and Data Security*, pages 508–526. Springer, 2019.

[300] István András Seres, László Gulyás, Dániel A Nagy, and Péter Burcsi. Topological Analysis of Bitcoins Lightning Network. In *Mathematical Research for Blockchain Economy*, pages 1–12. Springer, 2020.

[301] Sesha Kethineni, Ying Cao, and Cassandra Dodge. Use of Bitcoin in Darknet Markets: Examining Facilitative Factors on Bitcoin-Related Crimes. *American Journal of Criminal Justice*, 43(2):141–157, 2018.

[302] Fabian Maximilian Johannes Teichmann. Financing Yerrorism Yhrough Vryptocurrencies–A Danger for Europe? *Journal of Money Laundering Control*, 21(4): 513–519, 2018. https://doi.org/10.1108/JMLC-06-2017-0024.

[303] Dirk A Zetzsche, Ross P Buckley, Douglas W Arner, and Linus Föhr. The ICO Gold Rush: It's a Scam, It's a Bubble, It's a Super Challenge for Regulators. *University of Luxembourg Law Working Paper*, (11):17–83, 2017.

[304] Xiaoqi Li, Peng Jiang, Ting Chen, Xiapu Luo, and Qiaoyan Wen. A Survey on The Security of Blockchain Dystems. *Future Generation Computer Systems*, 107:841–853, 2020.

[305] Simone Porru, Andrea Pinna, Michele Marchesi, and Roberto Tonelli. Blockchain-Oriented Software Engineering: Challenges and New Directions. In *2017 IEEE/ACM 39th International Conference on Software Engineering Companion (ICSE-C)*, pages 169–171. IEEE, 2017.

[306] Rafael Pass, Lior Seeman, and Abhi Shelat. Analysis of the Blockchain Protocol in Asynchronous Networks. In *Annual International Conference on the Theory and Applications of Cryptographic Techniques*, pages 643–673. Springer, 2017.

[307] Savita Mohurle and Manisha Patil. A Brief Study of Wannacry Threat: Ransomware Attack 2017. *International Journal of Advanced Research in Computer Science*, 8(5), 2017.

[308] Keke Gai, Meikang Qiu, and Xiaotong Sun. A Survey on FinTech. *Journal of Network and Computer Applications*, 103:262–273, 2018.

[309] Germany Becomes First Country to Formally Recognize and Regulate Bitcoin;. *https://www.leaprate.com/news/germany-becomes-first-country-to-formally-recognize-and-regulate-bitcoin/ (accessed Jan 14, 2021).*

[310] China To Support Blockchain Development Under New Five-Year Plan. *https://www.ccn.com/china-support-blockchain-development-new-five-year-plan/ (accessed Jan 14, 2021).*

[311] Financial Conduct Authority. Regulatory sandbox. *https://www.fca.org.uk/publication/research/regulatory-sandbox.pdf (accessed Jan 14, 2021).*

[312] Michèle Finck. *Blockchain Regulation and Governance in Europe.* Cambridge University Press, 2018.

[313] The Inside Story of Mt. Gox, Bitcoin's $460 Million Disaster;. *https://www.wired.com/2014/03/bitcoin-exchange/ (accessed Jan 14, 2021).*

[314] Office of the Comptroller of the Currency (OCC). OCC Responsible Innovation Framework.

[315] Chinese Ministry of Industry and Information Technology. 2016 Chinese Blockchain Development. White Paper. *http://www.199it.com/archives/526865.html (accessed Jan 14, 2021).*, 2016.

[316] Distributed Ledger Technology: beyond block chain. *https://assets.publishing.service.gov.uk/government/uploads/ system/uploads/attachment_data/file/492972/gs-16-1-distributed-ledger-technology.pdf (accessed Jan 14, 2021).*

[317] Saqib Ali, Guojun Wang, Bebo White, and Komal Fatima. Libra Critique Towards Global Decentralized Financial System. In *International Conference on Smart City and Informatization*, pages 661–672. Springer, 2019.

[318] Z Yang, Decoding China's Cryptography Law. The Diplomat, 2019. https://thediplomat.com/2019/10/decoding-chinas-cryptography-law/

[319] Rosario Girasa. *Regulation of Cryptocurrencies and Blockchain Technologies: National and International Perspectives.* Springer, 2018.

[320] Gareth Peters, Efstathios Panayi, and Ariane Chapelle. Trends in Cryptocurrencies and Blockchain Technologies: A Monetary Theory and Regulation Perspective. *Journal of Financial Perspectives*, 3(3), 2015.

[321] Douglas W Arner, Janos Barberis, and Ross P Buckey. FinTech, RegTech, and the Reconceptualization of Financial Regulation. *Nw. J. Int'l L. & Bus.*, 37:371, 2016.

[322] Ringe, Wolf-Georg and Ruof, Christopher, A Regulatory Sandbox for Robo Advice (May 31, 2018). European Banking Institute Working Paper Series 2018 - no. 26, Available at SSRN: https://ssrn.com/abstract=3188828 or http://dx.doi.org/10.2139/ssrn.3188828

[323] China Has Banned ICOs. *https://techcrunch.com/2017/09/04/chinas-central-bank-has-banned-icos/ (accessed Jan 14, 2021).*

[324] Eric A. Fischer, Cybersecurity Issues and Challenges: In Brief, Congressional Research Service, 2016. Available on-line at: https://a51.nl/sites/default/files/pdf/R43831.pdf.

[325] Christopher C Chen. Regulatory Sandbox and InsurTech: A Preliminary Survey in Selected Countries. *Available at SSRN 3275929*, 2018.

[326] Fan Liao. Does China Need the Regulatory Sandbox? A Preliminary Analysis of Its Desirability as an Appropriate Mechanism for Regulating Fintech in China. In *Regulating FinTech in Asia*, pages 81–95. Springer, 2020.

[327] HaeOk Choi and KwangHo Lee. Micro-Operating Mechanism Approach for Regulatory Sandbox Policy Focused on Fintech. *Sustainability*, 12(19):8126, 2020.

[328] Luke G Thomas. The Case for A Federal Regulatory Sandbox for Fintech Companies. *NC Banking Inst.*, 22:257, 2018.

[329] Regulatory Sandbox for Fininal Technology. *https://www.hkma.gov.hk/gb_chi (accessed Jan 01, 2021).*

[330] PolicyPal becomes the First Start-Up to Graduate from the Mas Fintech Regulatory Sandbox. *https://fintechnews.sg/11127/insurtech/policypal-becomes-first-start-graduate-mas-fintech-regulatory-sandbox/ (accessed Jan 14, 2021).*

[331] Canada Extends Sandbox to ICOs, Impak Becomes Worlds First Regulated Token Sale. *https://www.trustnodes.com/2017/09/20/canada-extends-sandbox-icos-impak-becomes-worlds-first-regulated-token-sale (accessed Jan 14, 2021).*

[332] UK's FCA Wants to Promote A Global Regulatory Sandbox. *https://www.bbva.com/en/fca-assessing-feasibility-global-regulatory-sandbox/ (accessed Jan 14, 2021).*

[333] The GFIN is The International Network of Financial Regulators and Related Organisations Committed to Supporting Financial Innovation in The Best Interests of Consumers. *https://www.thegfin.com/ (accessed Jan 14, 2021).*

[334] Yhlas Sovbetov. Factors Influencing Cryptocurrency Prices: Evidence from Bitcoin, Ethereum, Dash, Litcoin, and Monero. *Journal of Economics and Financial Analysis*, 2(2):1–27, 2018.

[335] Shaen Corbet, Brian Lucey, and Larisa Yarovaya. Datestamping The Bitcoin and Ethereum bubbles. *Finance Research Letters*, 26:81–88, 2018.

[336] Walid Mensi, Khamis Hamed Al-Yahyaee, and Sang Hoon Kang. Structural Breaks and Double Long Memory of Cryptocurrency Prices: A Comparative Analysis from Bitcoin and Ethereum. *Finance Research Letters*, 29:222–230, 2019.

[337] M Poongodi, Ashutosh Sharma, V Vijayakumar, Vaibhav Bhardwaj, Abhinav Parkash Sharma, Razi Iqbal, and Rajiv Kumar. Prediction of The Price of Ethereum Blockchain Cryptocurrency in an Industrial Finance System. *Computers & Electrical Engineering*, 81:106527, 2020.

[338] Elie Bouri, Syed Jawad Hussain Shahzad, and David Roubaud. Co-explosivity in The Cryptocurrency Market. *Finance Research Letters*, 29:178–183, 2019.

[339] M Sy and S Morris. Pricing of Cryptocurrency-Use of Deep Learning and Recurrent Neural Networks Technology-Application to Bitcoin, Ethereum and Litecoin-Empirical Evidence. In *25th Annual Conference of the Multinational Finance Society*, pages 1–37. Global Business Publications, 2018.

[340] Kang Liu, Xiaoyu Qiu, Wuhui Chen, Xu Chen, and Zibin Zheng. Optimal Pricing Mechanism for Data Market in Blockchain-Enhanced Internet of Things. *IEEE Internet of Things Journal*, 6(6):9748–9761, 2019.

[341] Gianni Fenu, Lodovica Marchesi, Michele Marchesi, and Roberto Tonelli. The ICO Phenomenon and Its Relationships with Ethereum Smart Contract Environment. In *2018 International Workshop on Blockchain Oriented Software Engineering (IWBOSE)*, pages 26–32. IEEE, 2018.

[342] Christian Fisch. Initial Coin Offerings (ICOs) to Finance New Ventures. *Journal of Business Venturing*, 34(1):1–22, 2019.

[343] Weili Chen, Zibin Zheng, Jiahui Cui, and et al. Detecting Ponzi Schemes on Ethereum: Towards Healthier Blockchain Technology. In *Proceedings of the 2018 World Wide Web Conference*, pages 1409–1418, 2018.

[344] Fabio Massacci, Chan Nam Ngo, Jing Nie, Daniele Venturi, and Julian Williams. The Seconomics (Security-Economics) Vulnerabilities of Decentralized Autonomous Organizations. In *Cambridge International Workshop on Security Protocols*, pages 171–179. Springer, 2017.

[345] Sinclair Davidson, Primavera De Filippi, and Jason Potts. Economics of Blockchain. *Available at SSRN 2744751*, 2016.

[346] Sinclair Davidson, Primavera De Filippi, and Jason Potts. Disrupting Governance: The New Institutional Economics of Distributed Ledger Technology. *Available at SSRN 2811995*, 2016.

[347] Rainer Böhme, Nicolas Christin, Benjamin Edelman, and Tyler Moore. Bitcoin: Economics, Technology, and Governance. *Journal of economic Perspectives*, 29(2):213–38, 2015.

[348] Sergei Tikhomirov. Ethereum: State of Knowledge and Research Perspectives. In *International Symposium on Foundations and Practice of Security*, pages 206–221. Springer, 2017.

[349] Melanie Swan. Blockchain for Business: Next-Generation Enterprise Artificial Intelligence Systems. In *Advances in Computers*, volume 111, pages 121–162. Elsevier, 2018.

[350] Brett Scott. How Can Cryptocurrency and Blockchain Technology Play A Role in Building Social and Solidarity Finance? Technical report, UNRISD Working Paper, 2016.

[351] Alan T Norman. *Cryptocurrency Investing Bible*. CreateSpace Independent Publishing Platform, 2017.

[352] Orlenys López-Pintado, Luciano García-Bañuelos, Marlon Dumas, Ingo Weber, and Alexander Ponomarev. Caterpillar: A Business Process Execution Engine on The Ethereum Blockchain. *Software: Practice and Experience*, 49(7):1162–1193, 2019.

[353] Orlenys López-Pintado, Luciano García-Bañuelos, Marlon Dumas, and Ingo Weber. Caterpillar: A Blockchain-Based Business Process Management System. In *BPM (Demos)*, 2017.

[354] Stephan Haarmann, Kimon Batoulis, Adriatik Nikaj, and Mathias Weske. DMN Decision Execution on The Ethereum Blockchain. In *International Conference on Advanced Information Systems Engineering*, pages 327–341. Springer, 2018.

[355] Paul Rimba, An Binh Tran, Ingo Weber, Mark Staples, Alexander Ponomarev, and Xiwei Xu. Comparing Blockchain and Cloud Services for Business Process Execution. In *2017 IEEE International Conference on Software Architecture (ICSA)*, pages 257–260. IEEE, 2017.

[356] Anja Meironke, Tobias Seyffarth, and Johannes Damarowsky. Business Process Compliance and Blockchain.

[357] Luciano García-Bañuelos, Alexander Ponomarev, Marlon Dumas, and Ingo Weber. Optimized Execution of Business Processes on Blockchain. In *International Conference on Business Process Management*, pages 130–146. Springer, 2017.

[358] Aleksandr Kapitonov, Sergey Lonshakov, Aleksandr Krupenkin, and Ivan Berman. Blockchain-based Protocol of Autonomous Business Activity for Multi-Agent Systems Consisting of UAVs. In *2017 Workshop on Research, Education and Development of Unmanned Aerial Systems (RED-UAS)*, pages 84–89. IEEE, 2017.

[359] Claudio Di Ciccio, Alessio Cecconi, Jan Mendling, and et al. Blockchain-based Traceability of Inter-Organisational Business Processes. In *International Symposium on Business Modeling and Software Design*, pages 56–68. Springer, 2018.

[360] Ingo Weber, Xiwei Xu, Régis Riveret, and et al. Untrusted Business Process Monitoring and Execution Using Blockchain. In *International Conference on Business Process Management*, pages 329–347. Springer, 2016.

[361] Dang Hai Luong. The Ethereum Blockchain: Use Cases for Social Finance Applications. B.S. thesis, 2019.

[362] Alex Biryukov, Dmitry Khovratovich, and Sergei Tikhomirov. Findel: Secure Derivative Contracts for Ethereum. In *International Conference on Financial Cryptography and Data Security*, pages 453–467. Springer, 2017.

[363] Vladyslav Kopylash. An Ethereum-based Real Estate Application with Tampering-resilient Document Storage. *University of Tartu*, 2018.

[364] Beck, R., Avital, M., Rossi, M. et al. Blockchain Technology in Business and Information Systems Research. Bus Inf Syst Eng 59, 381–384 (2017). https://doi.org/10.1007/s12599-017-0505-1.

[365] Zhi Li, Hanyang Guo, Wai Ming Wang, Yijiang Guan, Ali Vatankhah Barenji, George Q Huang, Kevin S McFall, and Xin Chen. A blockchain and automl approach for open and automated customer service. *IEEE Transactions on Industrial Informatics*, 15(6):3642–3651, 2019.

[366] Ahmed Refaey, Karim Hammad, Sebastian Magierowski, and Ekram Hossain. A blockchain policy and charging control framework for roaming in cellular networks. *IEEE Network*, 2019, 34(3): 170–177.